MASTERING

STATISTICS
WITH YOUR MICROCOMPUTER

MACMILLAN MASTER SERiES

MASTERING
STATISTICS WITH YOUR
MICROCOMPUTER

CONALL BOYLE

MACMILLAN

First published 1986

Published by
MACMILLAN EDUCATION LTD
Houndmills, Basingstoke, Hampshire RG21 2XS
and London
Companies and representatives
throughout the world

Typeset by TecSet Ltd, Sutton, Surrey

Printed and bound in Great Britain by
Anchor Brendon Ltd, Tiptree, Essex

British Library Cataloguing in Publication Data
Boyle, Conall
Mastering statistics with your microcomputer. —
(Macmillan master series)
1. Statistics—Data processing
2. Microprocessors
I. Title
519.5'028'5404   QA276.4
ISBN 0-333-40917-5
ISBN 0-333-39172-1 Pbk
ISBN 0-333-39173-X Pbk export

To my boys Alex and Marcus
in the hope that their joy in finding out will never be diminished

CONTENTS

CONTENTS

PART III SUM UP AND EXPLAIN THE DATA

CONTENTS

ACKNOWLEDGEMENTS

I would like to thank many people for helping me to achieve this book.

John Bibby of the Open University for encouraging me to try EDA.

My colleagues at Birmingham Polytechnic, especially Bob Farmer for his encouragement, and Roger Ball for his tolerance; Andrew King and Gordon Kelly for their help.

The students on quantity surveying and estate management degree courses who so willingly tried out earlier drafts of this book, and from whom I learned so much. We lecturers must never forget that without our students we are nothing.

My wife Nesta who understands that doing this book is important to me.

I would also like to thank the Consumers' Association, publishers of *Which?*, for permission to use the diagram in Figure 7.4; and the editor of the *Guardian* for permission to reprint an extract in Chapter 12.

CONALL BOYLE

ACKNOWLEDGEMENTS

INTRODUCTION

This is a learning-by-doing book. With the aid of your microcomputer, it aims to introduce you to the delights of statistics. You will also learn a great deal about using your computer to produce results. In each chapter in Parts I–V there are a number of *Tasks* for you to do. Completing these tasks will enable you to master statistics.

INTRODUCTION

THE EQUIPMENT YOU WILL NEED FOR THESE LEARNING TASKS

- A microcomputer. Any sort will do so long as it can run BASIC programs.
- A cassette recorder attached to your micro for program and data recording (or a disk unit if you are fortunate).
- A supply of blank cassettes (or disks).
- A printer is *very* useful, but you can get by without one.

A method of recording your results – a ring binder or a notebook – is also needed. As you complete each task, you should write it up and file it away. This will give you a record of the progress you are making, and will be handy to refer to later.

THE PREVIOUS KNOWLEDGE YOU REQUIRE FOR THIS BOOK

- Mathematics – simple arithmetic is all you really need. This is a statistics course based on understanding, not algebra.
- Computer programming – since you have a computer, you should have learned a little BASIC by now. At least a smattering of BASIC is needed before you start on this course. Your instruction manual will fill in any gaps in your knowledge of programming.

ORGANISING FOR LEARNING

As this book is designed for self-learning, I think you will find it most helpful to get yourself organised. If you draw up a check-list of tasks, you can record when you start and finish each one. You can also set yourself target dates by which you hope to finish each chapter and part.

An *assignment* in this book could consist of a completed set of tasks from a single part. So your first assignment could be to complete the tasks in Part I.

I do hope you will enjoy yourself exploring and mastering statistics with the help of your microcomputer!

PART I

COLLECTING

THE DATA

This part is about collecting data. You will look at pencil and paper methods, and also use your computer to store the data.

DATA BASE

This book is about exploring data with the aid of your microcomputer. But before you can do any exploring on the computer, you need some data to work on. Here we concentrate on the *data*; the computing will start in the next chapter.

1.1 DATA: WHAT IS IT? WHERE CAN YOU FIND IT?

Data: that's something to do with numbers, isn't it? Well yes it is, but not just any numbers. In statistics we are concerned with finding out; discovering if the facts tell us a story. You can find the facts about a topic by collecting some data, usually as numbers, but it could also be in the form of descriptions. Data can be found in all sorts of places, quite often nearby. One thing which you may frequently have to hand is the daily newspaper. Papers are full of detailed information; let's have a look at some of it.

Remember this is a learning-by-doing text, so you are going to find out about data by looking for some. What follows is the first of the tasks which direct you along this self-learning path.

TASK 1.1 Find a source of data, for example a newspaper, and identify some examples of the information it contains. Note what are the topics for which data are given, what units are used, and how much data is available in each issue.

For this task you could make use of a newspaper, mainly because it is cheap and handy. Remember to record the results of your investigations in a notebook or ring binder. The entry could look like this:

DATA SOURCE: *Evening Mail,* Tues, 24 Sept. 1985.

Data topic	Units of measurement	No. of items per issue
1. Share prices	£ per share or p per share	500 approx.
2. Football teams	rankings 1st, 2nd, 3rd . . .	82 league teams
3. House prices	£ per house	700 +

Data as you can see is found everywhere in abundance. It is concerned with numbers in context, meaningful numbers. The data tells us something, it is information about a topic. Data can also include descriptions, like male/female or disagree/agree. We will be coming back to the different forms of measurement again in Chapter 4.

But what are we to make of all this data? Only rarely do 'the facts' speak for themselves, whatever the old saying might suggest. To make some sense of the data, to turn information into understanding, we have to explore further.

1.2 SETTING UP THE DATA BASE: HOUSE PRICES

Statistics is about real data in the real world. The best way to get a feel for real data is to collect some for yourself. In the next task, I am going to get you to collect a set of data, or 'create a data base' in computer jargon. This data base will be the test bed or practice ground for much of your exploration of statistics.

I have suggested the topic of house prices, because it is of widespread interest. You almost certainly live in some kind of dwelling. Maybe you pay rent for it. If you are not yet a home owner you will probably become one sooner or later. Many professionals are involved in housing. Surveyors, lawyers, architects, financiers are all directly involved in housing and house prices.

House prices also works well as a topic for data exploration. The data are cheap, handy and abundant. Information can easily be found for your own area, whether it is the suburb where you live or the town or stretch of countryside.

If, however, there is some other topic on which you are keen to collect data, please do so. I want you to feel a personal commitment to your own data, that your investigations are directed at a topic which interests you, and you find relevant. Indeed, this commitment to your own data is a vital part of your learning process.

TASK 1.2   Data on house prices:

What sources are available?
What is covered by each source? (areas, types)
Which source is best for your data base?

You will probably start with the aim of finding out about house prices in your own locality. But instead of just plumping for the first data source you come across, look around at all the possibilities. You may find that your first ideas on what is useful will change after you survey the full range of sources.

Possible sources for data on house prices might include:

(1) Daily newspapers, often with a special day for house sales.
(2) Weekly newspapers.
(3) Free advertising 'newspapers'.
(4) Estate agents' particulars.

For each source I suggest you record the following information in your notebook:

(1) Name and date of source, e.g. adverts in the *Evening Mail*, Friday, 13 Feb. 1984.
(2) What is the main area covered by the source?
(3) Who originates the information, i.e. who pays for it?
(4) How often do the adverts or other information appear? (Is the information up to date?)
(5) How many houses for sale can be found in one issue?

Having looked at the various sources of data available, you should make a decision: which source gives the most suitable information? Having settled that question, you can go ahead and collect the data.

TASK 1.3   Record the information for the data base.
Obtain at least 100 advertisements for houses for sale.
For each property record the following information:

asking price
number of bedrooms
type of garage

You should draw up some sort of 'coding sheet' to organise the results. It could look like this:

Record No.	Asking price	No. of bedrooms	Type of garage
1	£23 570	3	single
2	£56 250	5	double
3	£38 450	4	none

You must get at least 100 properties. This is needed so that you have a good size sample of data to work with. Once you get it on the computer in the next chapter, this amount of data can be easily handled. In comparison with the amounts of data that arise in commercial organisations and on statistical surveys, it is quite a small amount.

It is a good idea to cross-reference your data. This could be done by numbering each advertisement, and recording the number on the coding

To avoid this problem — cut your source ads from an old newspaper!

sheet. In this way you should be able to get back to the original source of the data. Hold on to your 'source' materials – you will almost certainly need to refer to them later.

On completion of this task, you should have a set of data, written onto your coding sheets, together with the original source material from which your data was drawn.

1.3 **LOOKING AHEAD**

Now that you have collected your data set, it might be helpful to look ahead. To show this I have produced the diagram in Fig. 1.1. (The numbers in the figure refer to chapters in the book.)

Fig 1.1

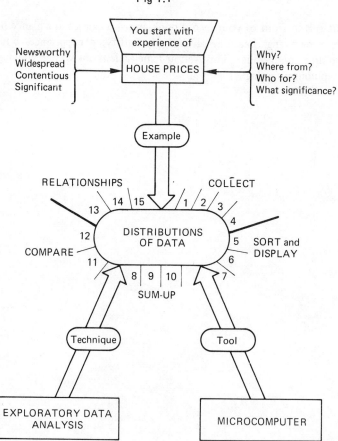

Figure 1.1 is an 'Advance Organiser'. There will be a number of these throughout this book to help you see the way the topics are set out. As this book is intended for self-learning, you might find it useful to keep an eye on your progress, to organise your own learning. This course is Task-driven — you learn by completing the tasks. So you could draw up an Action Time-table like this:

ACTION TIME-TABLE

Task no. and description	Started		Finished	
	target	actual	target	actual
1.1 Find source	Mon 5	Mon 5	Tue 6	Tue 6
1.2 House price sce	Tue 6	Wed 7	Wed 7	
1.3 Record data	Wed 7		Thu 8	

Put this in front of your record book and you can see at a glance how you are progressing. Learning is a deeply satisfying experience, but it also requires perseverance. This Action Time-table is one way in which you can keep yourself going!

CHAPTER 2

STORE DATA ON
YOUR COMPUTER

Now that you have collected your data, you need to make use of your computer to store it. In this chapter we get to grips with the computer for the first time. If your knowledge of BASIC is very patchy, then you might find some of the material a bit baffling. But I would expect if you have written even the simplest of programs, that you will take this chapter in your stride!

Of course the computer is not the only way of storing data. But once you get data in, get it into 'machine-readable' form, then it becomes very easy to do all kinds of tricks with it. After all it's what your electronic data processor (or computer as it is usually called) is for. And although you may think that 100 pieces of data is quite a lot, in surveys many more times this amount is encountered. The computer then becomes a vital tool in making sense of the data.

There are many ways of getting data into the computer. We are going to look at *four*:

(1) INPUT
(2) READ...DATA
(3) READ to an array
(4) READ/PRINT to a file

Getting the data into machine-readable form is invariably the most costly and time consuming part of the data analysis. Most texts skate over this bit as if it was trivial. It's because of the difficulties I know exist at this stage, that I am proposing you look at different ways of inputting data to the computer.

2.1 USE OF THE INPUT COMMAND

Many BASIC programs use the INPUT command, which stops the computer, and waits for the user to key in data and then press RETURN (or

ENTER). INPUT has advantages; you get your data in quickly, but it also has serious drawbacks:

(1) If you put in a wrong number, you can't easily correct it.
(2) Once entered and used the numbers are lost.

For these reasons I will not be suggesting that you use INPUT to enter your data.

2.2 USE OF READ . . . DATA

For our purposes the READ . . . DATA command is much better, even though it takes a bit longer. So the first task is to set up your data in DATA lines.

TASK 2.1   Set up your survey data on DATA lines, and READ it back as a check. Record the DATA lines on cassette (or disk).

On screen you could get something like:

```
1000 REM price in £, no of bedrms, garage

1010 REM garage coding 0-none, 1-single, 2-double, 3-parking

1020 DATA 25000, 2, 1

1030 DATA 35000, 3, 2

1040 DATA 29500, 3, 0
     .
     .
     .
```

and so on for 100 values.

Also as a check, you should read in and print back the data:

```
10 READ P, B, G

20 PRINT P, B, G

30 GO TO 10
```

If you don't quite follow what happened here, let me explain:
Line numbers; why start at 1000 and go up in steps of 10? Steps of 10 is a convention in BASIC. It allows you to insert extra bits of data later on. Starting at 1000 means that when you add in program lines, they can come before the data.

(You may be able to make use of a command like AUTO 1000,10 to generate line numbers automatically.)

Coding of garage: why not write 'SINGLE' 'DOUBLE'? One reason is that it is tedious and mistake-prone. Also computers prefer numbers; it makes programming easier if all data are numerical. It is a very good idea to have the coding scheme in a REM line. It is amazing how quickly you will forget such details. Don't rely on your memory.

Houses per line: why only one house per line? You could enter multiples:

1020 DATA 25000, 2, 1, 35000, 3, 2, 29500, 3, 0

but one house per line makes checking a lot easier. The big advantage of using the DATA lines is that you can use the normal editing and correcting facilities in BASIC. Adding new data, correcting existing lines, and deleting is all as simple as changing a line in a program. This is the real strength of the DATA command.

READ: This is a very simple way of getting the computer to look at the data. PRINT tells you what the computer has taken in. The aim of these lines is to check out the data for any mistakes, like missed commas, or mixed up lines.

ERROR AT LINE 10: OUT OF DATA

Did you get a message like this when you ran the program? It is always annoying when our machines tell us that we have blundered, but in this case the error is deliberate. Can you see why? In the next task we will overcome this little difficulty.

Once you are satisfied that the DATA are satisfactory, save onto tape or disk. Don't forget to label the tape (or disk); you will be making quite a lot of use of these data in the later tasks. As an additional safeguard, it would be very useful to list out the data using a printer, if you have one. In any event you should write down what you have done in your record book.

TASK 2.2 Count how many items of data there are in your set, i.e. how many houses you have recorded.

You load up the program and data from Task 2.1, and add the following lines:

    5 LET N = 0

    15 IF P < 0 THEN STOP

    17 LET N = N + 1

```
25 PRINT "                         COUNT =";N

9999 DATA -1, 0, 0
```

This is a conventional method of reading in data, using a signal value as a stop (in this case −1). It also can be used to count how many pieces of data you have got. No need to save this program, but make a note of it in your record book.

Putting your data on DATA lines has certainly got it into machine readable form, but has not strictly speaking got it into the computer. READ takes in one piece at a time, and then goes on to the next one. In the next section, we will tackle a method of getting all the data into the computer's memory, which has many advantages.

2.3 READ DATA INTO AN ARRAY (SUBSCRIPTED VARIABLE)

TASK 2.3   Read your data into an array, and use the program to set up an enquiry service

If you have not used arrays before please refer to your manual, and try out a simple example before attempting this one. Arrays and subscripted variables are usually treated as 'advanced topics' in BASIC, but for this course it is essential that you know how to handle them.

If you are familiar with arrays and subscripted variables, the following will not be too difficult:

```
10 DIM A(100,3)                   (I'm assuming count = 100)

20 REM 100 houses, 3 variables

22 REM now read in the data

30 FOR I = 1 TO 100

40    FOR J = 1 TO 3

50    READ A(I,J)

60    NEXT J

70 NEXT I

100 REM enquiry service

110 PRINT"FOR WHICH HOUSE NO. DO YOU REQUIRE DETAILS"
```

```
120 INPUT K

130 PRINT "PRICE = "; A(K,1)

140 PRINT A(K,2);"BEDROOMS"

200 GO TO 110
```

This is only my suggestion; you may wish to deal with it in some other way.

TWO ADDITIONS: this program is incomplete in two vital ways. Can you put these shortcomings right?

First: no print out on garage facilities. Can you arrange that the coding value 0, 1, 2, 3 is converted into words. So if 'G' is equal to 1, you print out 'single garage' (why, that's almost BASIC!).

Second; the enquiry service keeps coming back for more. Can you make it so that the enquirer can signal that he/she has no more requests (e.g. by inputting a negative value). And how would you get the program to deal with an enquiry for a house number greater than 100?

Don't forget to file away the program in your record book.

2.4 ENRICHMENT ACTIVITY - TWO EXTRA TASKS

That's the end of the essential tasks in this chapter. What follows are two extras, 'enrichment activity'. Both are well worth doing. If you cannot manage them now, you could come back to them later, but try to do them sometime.

TASK 2.4   Select a random sample of houses

To select a random sample we need to use the random number generator on the computer. This is usually of the form

RND(0)

which returns a random value in the interval 0 to 1. Again, if you have not come across this useful command, try it out, or refer to your manual. Details are highly variable, so check carefully.

The program to select a random sample is the same as Task 2.3, read into an array, but with the following amendments:

```
100 FOR K = 1 TO 5

120     I = INT(RND(0) * 100) + 1        (returns a whole no. in
                                          range 1 to 100)
130     PRINT "HOUSE NO "; I, "PRICE = £"; A(I,1)

200 NEXT K
```

Try running this program a couple of times. Do you get the same random sample? If so, you need to re-seed the random number generator, using a command like RANDOM, RANDOMISE, or RANDOMIZE. You may also like to improve the output; just giving the price is not very helpful. Why not print out the number of bedrooms and type of garage as well?

TASK 2.5   Read data on to a tape or disk file

In this task you are creating a real computer file. Because the details are so variable on different computers, I will not give a suggested program. Again, refer to your manual for details. This program is really no more difficult than Task 2.1, except that you write out the data to a device rather than to the screen. For example, in Microsoft BASIC you would enter (assuming that you have read the data into an array dimensioned A(100,3) as before):

```
10 FOR I = 1 TO 100

20    FOR J = 1 TO 3

30        PRINT£-1,A(I,J)          £-1 means device -1 i.e.
                                   tape (£ is # on many com-
40    NEXT J                       puters)

50 NEXT I
```

But this will be explained in your manual. If you have a disk drive, then forget about tape files - although tape is adequate for many purposes, floppy disk is far superior - but at a price of course. Here is an example of setting up a disk file in Microsoft BASIC (yet again the data is in an array A(I,J)).

```
100 OPEN "O",2,Z$

110 PRINT£2, A(I,J)

120 CLOSE 2
```

In Microsoft BASIC these lines mean:

"O" means the file is being set up for output, i.e. you are writing out from the computer to the disk,
Z$ is the name given to the file, and
2 is the device number.
CLOSE should be used when finished reading/writing.

CHAPTER 3

MANUAL

RECORDING

OF DATA

Manual recording of data? In this, the computer age? Well, the fact is that pencil and paper are often more effective than the computer. And when was the last time you had a fuse blow on a pencil? I am asking you to look at graphical means of recording data, first because they are useful in making sense of the data, in seeing what the picture is. But secondly, many of these manual techniques are so good, that it's worthwhile doing them on the computer as well. So although you will start with purely pencil and paper methods, later on you will get to use your computer to great effect.

3.1 TALLY MARK CHART

The first method is the tally mark chart. The data you have collected, on house prices range from a low point (the cheapest house), to a high value. The difference between these two extremes is called the RANGE.

RANGE = highest value − lowest value found in the set.

Because the set of values is distributed over the range, it is usually called the *distribution*. The range can be broken into a number of *intervals*. A tally mark chart is a way of counting how many items are to be found within each interval. It makes sense to break the range up into convenient intervals of, say £10 000.

The first stage in drawing your tally mark chart is to write out the 'stem' – the range of values broken into intervals. It would look like this:

Price £(x)	
10 000−	10 000 − means £10 000 and over
20 000−	and up to but
30 000−	not including £20 000
40 000−	
50 000−	
60 000−	

'Pencil and paper are often more effective than the computer.'

Once you have set up your scale, you can start logging the house prices. Say the first five prices were:

£25 365, £36 500, £49 995, £63 000, £25 000

On the chart they would be recorded as:

Price £(x)	
10 000–	
20 000–	/ /
30 000–	/
40 000–	/
50 000–	
60 000–	/

When a *fifth* value is recorded, draw a cross-bar, like this: ⊬⊬⊬. The final embellishment to your tally mark chart is to add a frequency column:

Price £(x)		f
10 000–		0
20 000	/ /	2
30 000–	/	1
40 000–	/	1
50 000–		0
60 000–	/	1
		$\Sigma f = 5$

where

x - what you measured (house prices here),
f - frequency in that interval,
Σ ('sigma') - sum of - add up all the items after the sign,
Σf - add up all the fs.

TASK 3.1   Plot a tally mark chart of the data on house prices.

When you have completed Task 3.1, don't forget to file it away in your record book. The benefits of drawing a tally chart are:

(1) It is a quick way of 'scratching down' the frequencies in each interval.
(2) It paints a picture; you can see the shape of the distribution develop before your very eyes.

There are snags of course. If your intervals are badly chosen, you finish up with a chart which is either too squashed, or too straggly. If you were using a computer to plot the tally chart, redrawing it would be very easy.

So,

TASK 3.2 Plot a tally mark chart using the computer. The program should allow the user to specify variable intervals. Use the program to try out the effect of variable intervals.

The first job is to set up the range and intervals for the chart. You might be clever and get the computer to choose these, but why not make use of that most powerful and flexible computer, your *brain*. Why not a few simple lines like:

```
10 INPUT "LOWER BOUND OF LOWEST INTERVAL";L
30 INPUT "INTERVAL LENGTH";IN
50 INPUT "MAXIMUM VALUE";M
```

Next you should read the data into an array, a repeat of what you did in Task 2.3. As before, there are 100 data values, which will be held in an array DIMensioned A(100,3). What you need to do now is to see if any of the values in A(I,J) lie within the first interval:

```
110 U = L + IN            that's the upper bound of the interval
120 FOR N = 1 TO 100
130  IF A(N,1) >= L AND A(N,1) < U THEN PRINT "/";
140 NEXT N
150 PRINT " "
160 L = L + IN
170 IF L < M THEN 110
```

This segment of program will effectively deal with the task. I should point out that this is not the most efficient method of getting the computer to draw a tally mark chart, but you need not worry about adding a few fractions of a second to the computer's processing time.

There are some improvements to the program that are just asking to be carried out:

Add on a title to the graph (ask for it in an INPUT)
Add a scale of values down the left-hand side
Add a count of how many values are found in each interval (this is most essential). A line should look like:

price £x	tally	f
25 000	//////////	10

If you count how many values are in each interval, you should also show the grand total (Σf).

Making use of your program:

Try out the program to see the effect of longer and shorter interval lengths. Some will give a picture which is too stretched out and straggly; some will be so compressed that the tallies 'wrap-round' the screen, making a complete mess of your picture. But when you do get the interval that gives a nice picture, just look how useful it is. At a glance you can see how the house prices are spread out, where the bulk of the prices are to be found, what the extremes are.

The tally mark chart is a very simple form of analysis, yet yields so much information. If you were to go no further than this task, you will have learned a most useful technique. (For my money, I would say *the* most useful bit of statistics you will ever learn.)

Don't forget to record the program and write up the results.

3.2 THE HISTOGRAM

If you replace the "/" with "*" you have almost got a *histogram*. The other difference is that the histogram has vertical bars. Histograms are a popular graphical method that has been around a long time. Figures 3.1 and 3.2 are an example of what they look like.

Fig 3.1   *Crude histogram using standard character set. Produced using the 'Minitab' statistics package*

```
--- HISTOGRAM OF 'AGES'

    MIDDLE OF      NUMBER OF
    INTERVAL       OBSERVATIONS
       20.             2        **
       25.             4        ****
       30.             4        ****
       35.             9        *********
       40.            11        ***********
       45.             7        *******
       50.             1        *
       55.             1        *
```

You can try to get the computer to produce a histogram with vertical bars, but if you have ever tried to produce this kind of graphical output, you will understand what a difficult problem it is. (In the tally mary chart life was made simple because we used horizontal bars.) So the next task is optional, for enthusiasts who wish to make use of their graphics facilities. No hints will be given – you are on your own!

Fig 3.2 *Conventional histogram from a lisa (Apple) microcomputer*

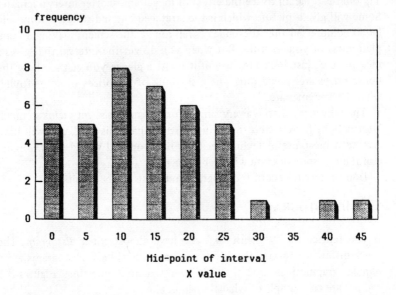

TASK 3.3 (Optional) Produce a computed histogram. Advanced versions would automatically choose intervals, axes, etc.

3.3 THE STEM AND LEAF PLOT

One shortcoming of the tally mark chart you may have noticed is that it throws away a lot of detail. The "/" just tells you that a value was found in the interval £20 to £30 thousand, giving no clue to its actual value. The stem and leaf plot rescues some or all of this detail, without too much cluttering up of your diagram. What you do is replace the simple "/" of the tally mark with one or more significant digits. For example:

£25 764 could be represented by:

Price £(x)

Stem	Leaf	
20 000−	5764	4-digit leaf
or		
20 000	57	2-digit leaf
or		
20 000	5	1-digit leaf

Note: this is *not* rounding-off.

A completed stem and leaf plot (2-digit recording) would look like this:

Stem	Leaf
£10 000s	£1000s and £100s
10 000	01 15
20 000	53 45 00
30 000	65 00 35 75
40 000	99 (means £49 900)
50 000	
60 000	30 (means £63 000)

The number of digits recorded depends on the amount of detail you want to record; usually one or at most two is quite enough. These values to the right of the stem are of course called the 'leaves', hence the name stem and leaf.

The stem and leaf plot was developed primarily as a pencil and paper technique by John Tukey, an American professor. It works just as well on the computer.

TASK 3.4   Plot, by computer, a stem and leaf diagram for house prices. The program should allow for different levels of recording, i.e. 1-digit, 4-digit.

This repeat of Task 3.2, plot a tally mark chart. The chief difference is that instead of plotting '/', you plot one or more digits. Can you get the computer to pick out the middle digits of a number? Here's how it can be done:

```
510 INPUT A                      A is a number like 25575
520  B = INT(A/10000)*10000      that gives the stem,
                                 when INterval = 1000
530 C = A - B                    C - the 'leaf' with all its digits
540 D = INT(C/1000)              D - single digit leaf
550 PRINT D: RETURN
```

So instead of PRINT"/"; it's GOSUB 510

Don't forget to record your program, and write up the results of this task.

3.4 FURTHER DEVELOPMENTS OF THE STEM AND LEAF PLOT

The version of the stem and leaf plot given above can be further elaborated for pencil and paper use.

Stems: significant digits only. Instead of writing

    20 000   for the stem,
    2   on its own would be enough.

It then is helpful to identify the units for the stem:

Stem	Leaf
10000s	1000s
2	0 1
3	5 4 0
4	6 0 3 7
5	9

Subdivided interval: To get a useful picture you don't always want to jump in unit steps in the stems. The stem intervals can be broken down as shown:

Full version	Brief version
Stem	Stem
10 000s	10 000s
20 000	2*
25 000	•
30 000	3*
35 000	•
40 000	4*

Where * is for leaves 0, 1, 2, 3, 4 and • represents 5, 6, 7, 8, 9.

or if a more finely divided scale is needed:

Full version	Brief version
Stem	Stem
10 000s	10 000s
20 000	2*
22 000	T
24 000	F
26 000	S
28 000	E
30 000	3*

TFSE stand for Two, Four, Six and Eight.

Brief versions are not so easy to program. They can also be confusing to read, so I would suggest that they be used only when vital.

3.5 SINGLE DIGIT OR CATEGORY PLOT: DOT BOX

With all this emphasis on plotting the distribution of your house prices, you may feel that we have been neglecting the other two bits of data, namely number of bedrooms, and type of garage. This section will remedy that neglect.

Of course, you could go ahead and plot a tally mark chart or a stem and leaf chart, but it would be rather cluttered and dull. Just imagine a line like this:

No. of bedrooms

   2   22222222   the leaf is 2 every time! (full version)

Instead of tallies, you can build up blocks of 10 this way

         2 . .   2 dots = 2 items

         2 : :   4 dots = 4 items

         2 :_: = 5 items

         2 ⊓̣   = 8 items

         2 ⊠̣   = 10 items

So a completed block ⊠̣ means 10, a part completed block ¦_: less than 10. It works just like the tally mark chart, but gives a much better picture. It is also safer than the tally mark chart, being less prone to misinterpretation. The result is called a Dot box chart

TASK 3.5 Plot (by hand) the distribution of bedroom numbers and garage facilities using the dot box chart

TYPES OF DATA AND SCALES OF MEASUREMENT

This is a short chapter, dealing with quite a philosophical topic. Remember the data on house prices? You collected three pieces of data for each house. Remind yourself what they were:

> Item 1 . . . which was in units of . . .
> Item 2 . . . which was in units of . . .
> Item 3 . . . which was in units of . . .

It is no coincidence that these three items represent the different types of data 'scales' in which values are measured.

(1) *Continuous data*   Data measured on a continuous numerical scale, such as time, temperature, weight.
(2) *Discrete data*   Data which are measured on a numerical scale, but in discrete, usually whole number, values. Examples are the number of wheels on a car, shoe sizes.
(3) *Classificatory data*   Data not on a number scale. They can be in order, such as exam results graded A, B, C, D. They can also be purely descriptive with no ranking, such as eye colour: blue, brown, grey, green.

TASK 4.1: This task requires thinking rather than doing, so you need to write out the reasons for your answer.

4.1.1.   Identify the most probable data type for each of the three items you measured.

4.1.2   Now suggest reasons why house prices could be either continuous or discrete.

4.1.3   Garage facilities come in more or less desirable classifications. Which was the most desirable garage category you found? Which the least desirable? Write down all of the types of garage facilities in order of desirability.

4.1.4   Using the ordered list of garage facilities from 4.1.3, and taking 'single garage' as scoring 1.0 units, arbitrarily assign 'scores' to all the other classifications. (Go on, use your judgement, and guess how much more 'double' is worth, and so on.)

Comment: Having completed the task, you are probably left wondering what you have learned. Measurement scales, it seems, are slippery undefinable things, which change depending on how you look at them. To a great extent that is true, but that's what things are like in the real world.

You may have noticed that the scales used in this chapter are a bit like the 'variable types' in BASIC. You recall these:

In BASIC	*Data scale type*
Integer (whole number only)	. . . .
Real (numbers with decimals)	. . . .
String (words, letters)	. . . .

No doubt you can fill in the 'data type' corresponding to the BASIC variable types. (There is also double precision in BASIC, but that is the same as Real, except with twice as many digits.)

Statistics had its origin in physical (pure) science. Men of science value greatly their objectivity; 'if you can measure something then you can say something about it', said the great Lord Kelvin. Objective measurement is the foundation of science. Scales of length, temperature, weight are rigorously defined.

Compare this with the measurement scales of social science. Attitude measurement is based on responses like 'agree', 'agree strongly', 'disagree', etc. By no stretch of the imagination can such scales be called precise. Scientists and engineers often pour scorn on sociologists for this very lack of measurable precision.

So should we restrict ourselves to continuous measurement scales only (plus perhaps the odd discrete scale) on the grounds that this is the only true, objective scientific measurement? Human interest does not stop at 'scientifically' measured items. We would cut ourselves off from vast areas of investigation if we did so.

But do not be taken in by the 'objective' nature of scientific measurement. Remember that all measured values are only approximate. Temperature can probably be measured to the nearest degree on an ordinary thermometer. So this is really a discrete scale. But is $60°$ twice as hot as $30°$ (on any scale, Celsius or Fahrenheit)?

From all of this I hope you can agree that:

ALL SCALES OF MEASUREMENT ARE ARBITRARY

Some scales give items in order, some do not. But a scale of millimetres and a scale of 'agree–disagree' are equally valid.

(I did warn you that this was going to be a philosophical chapter. It has touched on a controversial area, which is not generally agreed.)

PART II
SORT AND DISPLAY
THE DATA

In this part you will begin to explore your data, sorting them into order, and finding significant points along the data.

SORTING DATA INTO ORDER

A data set which has been sorted into order is a very useful tool for exploring. Just putting the houses in price order will give you a lot of insight into the structure of prices. You can do this sorting into order by hand, using the stem and leaf plot that you worked out previously in Task 3.4. For instance, if the full digit version of the interval £22 000 (meaning £22 000 and over, and up to £24 000) showed:

£22 000– 3500,<u>2950</u>,<u>2600</u>,3000,<u>2375</u>,<u>2000</u>,3000,3900,3700,<u>2500</u>

In order: 2000, 2375, 2500, 2600, 2950, etc.

1st   2nd   3rd   4th   5th   etc.

So long as the number of values in the interval is not excessive, then converting to an ordered list is straightforward. Start with the lowest value, cross it off (I've underlined here), then look for the next lowest, and so on. I've done this for the first five values; can you complete the list?

This is quite a slow process, so for the next task, restrict yourself to a section of the data (you can make life easier by rerunning the program from Task 3.4, with narrow intervals.)

TASK 5.1   From your house price data set, select the 25 most expensive houses, and manually sort them into price order.

(Although this is a 'manual' method, you did get a little help from your friend the computer; a totally manual method would be very time consuming. The expensive end of the range is usually the most interesting, but feel free to sort any selection, such as the cheapest 25 houses.)

Ordered data (or 'ranked' data as it is sometimes called) helps you to understand what sort of values you have got. This sorting procedure will be used in further analysis later. So we will stay with sorting, and develop fully computerised methods for producing an ordered set of data.

One method of sorting using the computer could be quite similar to the manual method you have just used. All you would need to do is narrow the interval to units of £1. But this method has serious drawbacks, and is not practical. To test this, try the following experiment:

TASK 5.2   Rerun Task 3.4 with an interval of £1, in an attempt to sort the data. Note how long it takes, and whether it 'works' (i.e. produces a sorted list).

So what did you discover?
Did you find that the program took an unacceptably long time to produce the results? Even with a machine code* program or compiled* BASIC, the method is just too slow. Remember also this list has only 100 entries – in many applications, much longer lists are encountered. You may not have noticed the 'two-at-the-same-value' problem. An example of this was found in the data exercise at the start of this chapter. There were two houses at £23 000, which came 6th and 7th (or is it equal 6th?). A proper sorting program should cope with any number of equal entries. The program in Task 5.2 did not adequately deal with this.

A sorting program has to be effective – to sort the data set effectively into order. It must be reasonably efficient – do the job within an acceptable time-span. The quest for efficient sorting algorithms* is a major endeavour of computer science; in the next task you will look at a program which is simple yet efficient.

TASK 5.3   Sort your house price data into price order using the 'selection sort, technique described in the next section.

Details and comparisons of sorting methods can be found in Lee and Lee, *Statistics and Computer Methods in BASIC* (Van Nostrand Reinhold, 1982). They list the following:

   Bubble sort
   Insertion sort
   Selection sort
   Address sort
   Shell sort

with the shell and address sort the winners in most cases. If you use a stan-

---
* Explanation of technical terms:
   Machine code is the low level language, which is close to the method of operating of the computer. It is many times faster than BASIC.
   Compiled BASIC overcomes some of the slowness of BASIC by translating the instructions into machine code.
   An *algorithm* is a formula or method for solving a problem.

dard computer package, you would hope that an efficient sorting algorithm has been used. But here, for learning purposes, we are going to look at one of the simpler methods – the selection sort, which is perfectly adequate for the task in hand.

The selection sort is a 'search-and-swap' technique. Let's assume that your 100 house prices are in an array declared as

    DIM A(100).

(1) Search through the data for the highest value, noting where you found it;
(2) Swap this highest value into the top slot in the array;
(3) Repeat for the second highest; swap into the second slot in the array.

By a series of passes the ordered set of data is achieved, each pass taking less effort than the previous one. This is illustrated in Fig. 5.1.

Fig 5.1

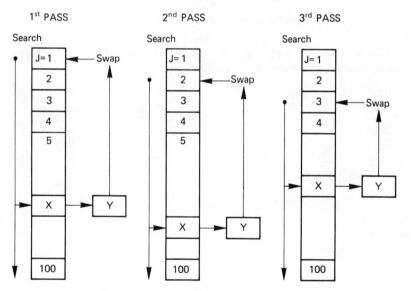

The program would look like this:

```
REM the data - 100 houses - is in A(I)
100 N = 100
110 FOR J = 1 TO N                          N passes
120    MAX = 0
130       FOR I = J TO N                    search
140          IF A(I) >= MAX THEN X = I       note place
150          IF A(I) >= MAX THEN MAX = A(I)  note value
160       NEXT I
```

```
170   Y = A(X): A(X) = A(J): A(J) = Y          swap A(X)<>A(J)
180 NEXT J
```

(Try out the program with a small set of data before sorting the full 100 values. Compare the time the final selection sort program takes with the crude and ineffective method you rejected in Task 5.2.)

Note that in this task you have sorted into descending order, that is you start with the biggest, then the next biggest, and so on. If you change the $>$ to a $<$ in both lines 140 and 150, you can sort into ascending order.

TASK 5.4   Sort your house price data set into ascending order.

This ordered set of data is going to be used quite a lot from now on, so it is a good idea to store it in a tape or disk file. In future, when you need to examine a sorted set of data, you will then have the option: (a) to read the unsorted data from DATA lines, and then sort it into order, or (b) to read sorted data direct from a file.

Remember you read data onto a file in Task 2.5.

TASK 5.5   Having sorted all the data on house prices into price order, store this ordered set on a file (tape or disk).

CHAPTER 6

FIND THE HINGES

One advantage of the ordered set of price data is the possibility of producing stem and leaf plots with the leaves in order. In Task 3.4, you produced a stem and leaf plot, by computer. With the ordered set of data it will be quite simple to repeat the plot, but this time with the leaves listed out in order. Keep the level of recording simple; wide intervals and single-digit leaves.

TASK 6.1 Produce a stem and leaf plot by computer, with the leaves in order.

6.1 COUNTS UP AND DOWN

From your ordered stem and leaf plot you can also add on counts of how many values were found in each interval, and cumulative counts, both upwards and downwards, of how many have been found so far.

To show what is intended, consider this set of examination marks.

$$45, 51, 55, 23, 65, 72, 78, 64, 61, 40, 39$$

The stem and leaf plot looks like this:

Examination marks:

Stem 10s	Leaf 1s	Count
2	3	1
3	9	1
4	0 5	2
5	1 5	2
6	1 4 5	3
7	2 8	2
		11 (total)

The cumulative count, that is the count of how many have been found so far, can be added to the plot as shown below:

Examination marks:

Stem	Leaf	Count	Count-up	Count-down
10s	1s			
2	3	1	1	11
3	9	1	2	10
4	0 5	2	4	9
5	1 5	2	6	7
6	1 4 5	3	9	5
7	2 8	2	11	2
		11		

Interpreting the meaning of the figures in the 'count-up' and 'count-down' columns requires some care. These values have an 'up-to' meaning: for instance, 4 in count-up means that 4 values were found up to the end of the 40s, i.e. which were less than 50. Similarly, the 7 in the count-down column means that 7 values were found down to the bottom of the 50s, i.e. which were greater than 49.

TASK 6.2   Enhance the plot of Task 5.1 with counts as shown above. Computer or manual solution is acceptable.

6.2   HINGES: THE MARKER POSTS ALONG THE DATA

So far we have collected and sorted the data. Now we are going to take things a stage further by picking out values along the data which help us to understand what it is saying. These are called the 'hinges' in the data.

Returning to the set of 11 exam. marks:

$$45, 51, 55, 23, 65, 72, 78, 64, 61, 40, 39$$

Sorting them into order:

```
Rank:    1   2   3   4   5   6   7   8   9  10  11  (bottom up)
        23, 39, 40, 45, 51, 55, 61, 64, 65, 72, 78
        11  10   9   8   7   6   5   4   3   2   1  (top down)
                             Middle
```

There are 11 values in all, ranging from 23 to 78. These two extremes are our first important markers. The middle of the row is the sixth value, which breaks the set into two equal halves. This can be represented as shown in Fig. 6.1.

Fig 6.1 *hinge point in the data: median M*_d

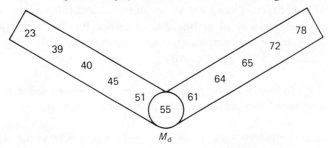

55 is the 'hinge' for the data, the point in the middle around which it bends, or is balanced. We can add in another pair of 'hinges', as shown in Fig. 6.2.

Fig 6.2 *hinge points in the data. Quartiles Q*ₗ *and Q*ᵤ*. Upper and lower points U and L*

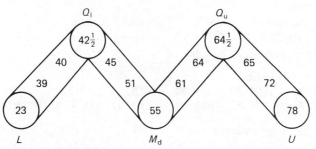

This time the set of ordered numbers does not break nicely, and we are left with a 'hinge' between 40 and 45 at the lower end, and 64 and 65 at the upper end. The obvious thing to do is invent a number where the hinge should be, ('interpolate') at a point halfway between the two.

$$\text{So } (40 + 45)/2 = 42.5 \quad \text{and} \quad (64 + 65)/2 = 64.5.$$

It is high time we labelled these hinge points and defined exactly where they are to be found.

Extremes The maximum and minimum points. The difference between these two points is called the *range*. There is no problem in locating these points if you have an ordered set of data. L – the lower extreme, U – the upper extreme.

Median The middle point of the distribution. If N (the number of data points in the set) is *odd*, then there is a middle point. If N is *even*, take the average of the two values nearest the middle, M_d the median.

Quartiles The values halfway between the extremes and the median. (You could say they are the 'quarter' points, hence the name.) Depending on how many values there are in your set, the quartiles will be at an exact point, or in between. Listing out all the numbers, and folding them along their hinges is one way of getting these quartiles. But this is tedious, so we are going to get the computer to use a formula to spot them. Q_l – the lower quartile, Q_u – the upper quartile.

TASK 6.3   Calculate all five hinge points on your house price data. Use the ordered set of data, and interpolate points where necessary.

Calculating the extremes, L and U, presents no difficulty; the median and quartiles need a bit more explanation:

If N is the number of values in the data set, then:

For the *median*: N_d calculate $(N + 1)/2$
    if result is a whole number use it
    otherwise take the average of the values either side
For example:
    if $N = 9$, then $(9 + 1)/2 = 5$,   so take 5th value
    if $N = 10$, then $(10 + 1)/2 = 5.5$,   so take average of 5th and 6th.
    In BASIC this would be achieved by

```
 90 M = (N+1)/2
100 IF M =  INT(M) THEN MD = A(M)
110 IF M <> INT(M) THEN MD = (A(INT(M)) + A(INT(M)+1))/2
```

For the *quartiles*: calculate $(N + 1)/4$ for Q_l;
          and $3(N + 1)/4$ for Q_u
    if the result is a whole number use it;
    otherwise take the average of values either side.

Purists may object that we should be using quarters as well as halves; that is correct, but it is an unnecessary elaboration. Simple averaging works all right.
    For example:

    if $N = 7$, then $(9 + 1)/4 = 2$ Q_l is the 2nd value
                              Q_u is the 6th value
    if $N = 9$, then $(10 + 1)/4 = 2.25$;
                              Q_l is the average of the 2nd and 3rd value
                              Q_u is the average of the 6th and 7th value

For programming, the method of calculation is identical to that for the median.

6.3 5-NUMBER SUMMARY

In the last section, you found the five hinge points, namely the two extremes, the median and the two quartiles. We need a standard method of presenting these results. To return to the exam. marks example, we found the following hinge points:

$$\begin{array}{lll} \text{extremes:} & L\text{:} & \text{value} = 23 \\ & U\text{:} & \text{value} = 78 \\ \text{median:} & M_d\text{:} & \text{value} = 55 \\ \text{quartiles:} & Q_1\text{:} & \text{value} = (40+45)/2 = 42.5 \\ & Q_u\text{:} & \text{value} = (64+65)/2 = 64.5 \end{array}$$

These hinge points can be shown schematically like this:

Examination marks				$N = 11$
$L = 23$	$Q_1 = 42.5$	$M_d = 55$	$Q_u = 64.5$	$U = 78$

TASK 6.4 Write out the results of the previous task in 5-number summary form.

We will be returning to these 5-number summaries, and making further use of them in Chapter 7.

6.4 DISCRETE AND CLASSIFICATORY DATA – SUMMARIES

This section deals with an important and useful topic – how to derive 5-point summaries for discrete and classificatory data. It is a little more complicated than previous sections, so if you get bogged down, press on with the next section, and return to this topic later. (I should also add that this section is my own interpretation of Tukey's EDA (exploratory data analysis) – I have found it necessary to extend the idea of hinge points to discrete data, because Tukey deals only with continuous, scientific values.)

The method described in the last section works well with continuous data. But in many applications, especially in business, the data values are either discrete of classificatory. If statistics cannot deal with these then it fails to cope adequately with the realities. The sort of problem that arises with discrete and classificatory data can be shown by the following example:

Exam grade	Leaf	N	Up	Down
A	AAA	3	3	24
B	BBBBBB	6	9	21
C	CCCCCCCCCCCCC	13	22	15
D	DD	2	24	2
E		0	24	0

24

Now you could draw up a 5-number summary:

Exam. grades				$N = 24$
L = D	Q_l = C	M_d = C	Q_u = C	U = A

But this does not tell us very much. The trouble arises with category C. To be told that the median and both quartiles are C, tells little about the scoring overall. Another set of students could turn in exactly the same 5-number summary, despite having scored a lot worse, getting a lot more Ds andEs. To extract more meaningful information from this summary, we need to refine the technique.

First, let us change the letter codes into number scores:

$$A - 5, \ B - 4, \ C - 3, \ D - 2, \ E - 1 \ \text{would do.}$$

Now let us extend this scale to include decimal values. C is no longer just 3. It can represent any score between 2.5 and 3.5. (Although the exam. victims will all be given a single letter code, on the marking, some will be only just C, some a very good C, most in between. So the subtle graduations are fully justified.)

Rewriting the results as scores:

exam. score	Leaf	N	Up	Down
5	555	3	3	24
4	444444	6	9	21
3	3333333333333	13	22	15
2	22	2	24	2
1		0	24	0
		24		

Fig 6.3 *twenty-four candidates lined up behind their scores*

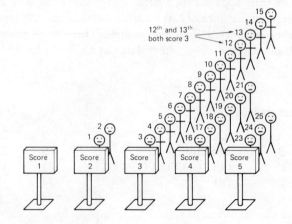

Since there are 24 values, the median is to be found at the 12.5th point. If all the individuals were lined up in order behind their scores, we would get a picture like that shown in Fig. 6.3.

Although it is quite correct to say that the median is 3 because that is the score of the middle pair, again it is not very informative. A median of 3 could be found with many different results. Imagine instead, that the candidates are strung out along the scale like starlings on a telephone wire like this:

To spread out our candidates, first 'enhance' the scale by filling in points in between. The boundaries between intervals now fall at 1.5, 2.5, 3.5, with the outer limits at 5.5 and 0.5.

Score |. . . . |. . . . |. . . .|. . . .|. . . .|. . . .|. . . .|. . . .|. . . .|. . . .|
          1.0   2.0   3.0   4.0   5.0

To see this in more detail, consider the three candidates who scored an A or 5 on the number scale. They fit into the interval 4.5 to 5.5, as shown. You can think of it like 3 starlings sharing 1 unit of space, so the space between is 1/3 = 0.333, or 1 standard bird-space. That leaves half a bird-space at either end, as shown in Fig. 6.4.

Fig 6.4 *three candidates scoring 5 spaced out in the interval 4.5-5.5*

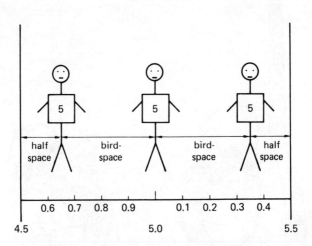

Space out the candidates with one 'bird-space' between them, and half a space at either end of the interval. To get to the 12th and 13th candidates:

they are both in the 2.5–3.5 interval
there are 13 candidates in this interval

giving

bird-space: 1/13 = 0.077
half bird-space: 0.077/2 = 0.038

In this interval the first candidate is the 3rd along, which is shown in Fig. 6.5.

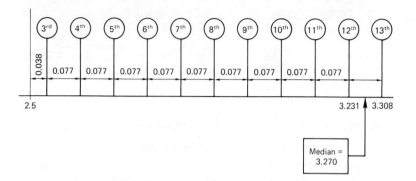

Fig 6.5 *median found between 12th and 13th candidates*

The other hinge points, i.e. the quartiles and the extremes are worked out in the same way. Spreading out discrete values evenly in the interval is not the only way you could cope with 'piled up' data. You could argue that the values might best be shown heaped up wards the middle of the distribution. Another suggestion is that the values be 'jittered' - randomly spread about in the interval. But remember (in Chapter 4,) you discovered that 'all scales are arbitrary' - spreading values evenly is simple and works well.

TASK 6.5 Calculate the exact hinge points for numbers of bedrooms in your house price data set.

This task can be undertaken with a very short program, which is given in full below:

```
5 REM *** N - NO. OF VALUES      MD - MEDIAN
6 REM *** L - LOWER EXTREME       U - UPPER EXTREME
7 REM *** QL - LOWER QUARTILE     QU - UPPER QUARTILE
10 INPUT "TOTAL NUMBER OF VALUES";N
20 PRINT "MEDIAN AT "; (N+1)/2
25      GOSUB 60
30 PRINT "LOWER QUARTILE AT "; (N+1)/4
35      GOSUB 60
40 PRINT "UPPER QUARTILE AT "; 3*(N+1)/4
45      GOSUB 60
50 PRINT "LOWER EXTREME AT   "; 1
51      GOSUB 60
55 PRINT "UPPER EXTREME AT   "; N
56      GOSUB 60
```

```
60 INPUT "FREQUENCY AT HINGE POINT "; HF
70 INPUT "INTEGER VALUE OF X IN THE INTERVAL "; I
80 INPUT "COUNT UP AT LOWER BOUND OF INTERVAL "; LB
90 INPUT "FREQUENCY IN THE HINGE INTERVAL"; K

150 V =  (HF - LB)*(1/K)  -  (1/K/2)  +  (I - 1/2)
160 PRINT "VALUE ON X SCALE AT HINGE "; V
170 RETURN
```

The key line is 150, which calculates the value V as:

$$V = (\text{bird-space}) \times (\text{no. of values to hinge})$$
$$- (\text{half a bird-space})$$
$$+ (\text{the integer value} - \text{a half, i.e. the lower bound})$$

This program makes full use of your human abilities to spot when to round-off; in the appendix to this chapter, I give a full listing of a program which will identify all five hinge points in discrete data which has been spread out.

6.5 WHAT DO HINGES TELL US ABOUT THE DATA?

Remember what we are doing – exploring the data. The five marker points L, Q_1, M_d, Q_u, U remind us of the spread of values in the data. If your median house price was £37 500, then you know that half the houses cost more, half less than this amount. The bottom quarter (price less than Q_1) indicates the cheaper end of the market. Interpreting the meaning of the hinge values in discrete data is not so obvious, so I will leave you to think about it.

TASK 6.6 Discussion *'The median number of rooms per house is 2.77', is the sort of figure you might find in the task. Do you feel that 0.77 of a room is a ridiculous quantity; or with the correct interpretation might it be a useful value? Discuss with others, or write out your own reasoning.*

6.6 APPENDIX: PROGRAM TO CALCULATE HINGE POINTS IN DISCRETE DATA

This program (Fig. 6.6) has been developed to deal with data in the form of scores 1, 2, 3, 4, 5. The program would have to be modified to cope with other scoring systems.

Fig 6.6

```
10 PRINT"PROG TO CALCULATE HINGE POINTS IN DISCRETE DATA"
11 PRINT"FOR DATA SCORED 1, 2, 3, 4, 5 ONLY"
17 FORO=1TO5:C(O)=0:D(O)=0:NEXTO
18 N= 0
20 REM **** READ IN THE DATA INTO C(I) EG ALL 3´S INTO C(3)
30 READ I
35 IF I < 0 THEN 60
37 N= N+1
40 C(I)=C(I)+1
50 GOTO 30
59 REM **** PRINT OUT SCORE RESULTS WITH COUNTS
60 D(1)=C(1)
61 LPRINT" "
63 LPRINT" SCORE         COUNT           COUNT-UP"
65 LPRINT 1,C(1),D(1)
70 FORI=2TO5:D(I)=D(I-1)+C(I)
80 LPRINT I,C(I),D(I)
90 NEXT I
100 GOTO 300
105 REM ****  SUBROUTINE TO SELECT INTERVAL FOR HINGES
110 S=0
120 FORI=1TO5
130 S=S+C(I)
140 IFS>=MN THEN 200
150 NEXTI
160 STOP
200 V=I-.5+(-.5+MN-D(I)+C(I))*(1/C(I))
205 V=INT(V*10+.5)/10
210 RETURN
300 REM **** CALC HINGE POINT FRQUENCY  MN, AND FIND VALUE V
310 MN=(N+1)/2
320 GOSUB110
340 MD=V
400 REM LOWER
410 MN=1
420 GOSUB110
430 L=V
500 REM U
510 MN=N
520 GOSUB 110
530 U=V
600 MN=.75*(N+1)
610 GOSUB110
620 QU=V
700 MN=.25*(N+1)
710 GOSUB110
720 QL=V
730 LPRINT" "
734 LPRINT"DISCRETE SCORES"TAB(55)"N=";N
735 LPRINT"-------------------------------------------"
736 LPRINT"L=";L,"QL=";QL,"MD=";MD,"QU=";QU,"U=";U
1000 DATA 2,5,2,2,3,5,5,2,2,1,5,5,1,5,1,3,4,5,5,4,4,4,-1
```

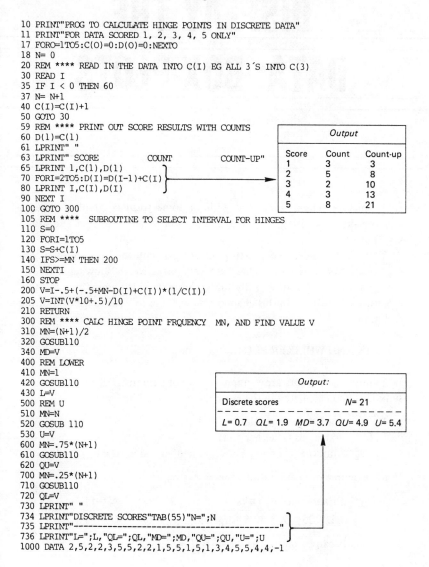

Output

Score	Count	Count-up
1	3	3
2	5	8
3	2	10
4	3	13
5	8	21

Output:

Discrete scores			$N= 21$	
$L= 0.7$	$QL= 1.9$	$MD= 3.7$	$QU= 4.9$	$U= 5.4$

DISPLAY THE
DATA: BOX-PLOTS

Every picture tells a story, and without some form of graph your data will remain uninspiring. 'There is no excuse for failing to plot and look', says Tukey.

You have done some plotting already, of course. In Chapter 2 you drew stem and leaf plots. These are very effective in showing what the data is saying, at a glance. Tally mark charts and histograms also give you a picture of the data. In this chapter you will add to your range of plotting techniques.

7.1 BOX AND WHISKER PLOTS

Box and whisker plots are a graphical way of demonstrating the 5-number summary you calculated in the last chapter. Recall the five points:

M_d median (in the middle),
L, U extremes (at either end),
Q_l, Q_u quartiles (halfway between median and extreme).

In the example of exam. results, the points were:

Examination marks				$N = 11$
L = 23	Q_l = 42.5	M_d = 55	Q_u = 64.5	U = 78

These points can be plotted as shown in Fig. 7.1.

Fig 7.1 *box-plot drawn by hand*

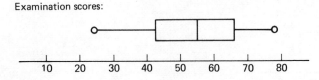

Examination scores:

10	20	30	40	50	60	70	80

'Box and whisker plot' (usually shortened to 'box-plot') is one of those made-up words which may tell you what the thing looks like, but gives you little clue as to its purpose. But the name box-plot has stuck, so we had all better get used to it! In the example in Fig. 7.1, the plot is drawn horizontally. You can also draw box-plots vertically, but drawing the box-plot horizontally is the more popular method. It is also much easier for computerised plotting, and horizontal plots will be used from now on.

TASK 7.1   Draw a box-plot (by hand) for house price data

Drawing the box-plots with the aid of the computer is the next stage. Ideally, you should use the special graphics characters on your computer, to get a smart diagram. Fig. 7.2 shows what can be done.

Fig. 7.2   *box plot using dot-matrix printer graphics*

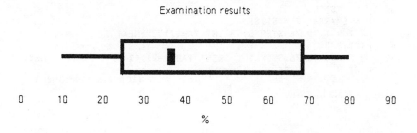

Because of the great variation in graphics between computers, it is usual to opt for a crude box-plot based on the standard characters available on all machines. For programming, it is much easier to plot horizontally. An example from the 'Minitab' statistics package is shown in Fig. 7.3.

Fig 7.3   *box plot using standard character set (produced using the 'Minitab' statistics package)*

TASK 7.2 Using computer graphics, draw a box-plot for house prices and numbers of bedrooms. (Input the five number summaries found in the last chapter.

Hints for Task 7.2:

Plotting a scale of values For this program, the main problem is deciding the ratio between actual values and VDU screen values. If your screen-width is 80 characters, plotting exam. results in the range 23 - 78 is a straight one-to-one operation. But if you plot house prices, then you need to scale down the values. For example:

If house prices range from £10 000 to £150 000, and you have 80 columns available, then

$$(150\,000 - 10\,000)/80 = 1750 \qquad \text{say £2000}$$

In the program, this could translate as:

```
10 INPUT "SCALE FACTOR"; SF      1000
20 INPUT "MINIMUM"; SM           10000
30 INPUT "STEPS"; ST             2000
35 SM = SM/SF: ST = ST/SF
40 FOR I = 1 TO 76 STEP 5: PRINT TAB(I)".":NEXT I
45 PRINT " "
50 FOR I = 1 TO 76 STEP 10: PRINT TAB(I-2) (I-1)*ST + SM : NEXT I
60 PRINT "   x £";SF
```
                                          (Input is underlined)

giving a scale like this:

```
   .     .     .     .     .     .     .     .     .     .     .     .     .     .
  10        30        50        70        90       110       130
   x £1000
```

Plotting the box and whisker Adding on the plot requires reading in L, U, QL, QU and MD. These values would be scaled down by 1000; if MD was £57 536, this would become 58. The top line of the box-plot spreads from QL to QU, and could be printed out as:

```
150 FOR I = QL/ST TO QU/ST: PRINT TAB(I)"-";: NEXT I
```

For single points:

```
180 PRINT TAB(MD/ST)"+";
```

7.2 HINTS FOR PLOTTERS

As the box-plot is the most important graphical presentation used in this book, a few general comments on presenting your results in a graphical form would be useful.

1. Plotting is always worthwhile. You may feel that the extra time it takes to plot could be spent in other ways, or that the facts on their

own are sufficient. Untrue! The majority of readers will skim over dull columns of figures; a picture will arrest the attention. A plot helps you to *see* what you might not have expected. Numbers alone seldom tell a story.

2. If plotting is worthwhile, good plotting is even better. Take a pride in your plots - they are your 'face' to the world. So a clean, well drawn, clearly annotated picture should always be the aim.

3. Use colour. Colour in your box and whisker plots. This improves the impact. It also helps the reader, especially if you use some form of colour coding to distinguish different features.

4. *Don't* present your results on graph paper. The grid lines on graph paper can obscure the meaning of a plot. The scales clutter things up. They also tend to make it untidy. If you want to plot accurately, make use of tracing paper, or acetate sheets. Place these over the graph paper which has the scales drawn out in full. Plot your picture, together with a skeleton scale.

5. Do you include zero or not? In some cases, it is vital. If you are comparing two sets of data in order to discover differences, then a plot down to zero is unnecessary, even misleading. On the occasions when it is the absolute size of the data values that matters, include the zero. In general though, avoid zero unless leaving it out would mislead.

6. If you have no way of accurately drawing box-plots or graphs, don't let that put you off. Plotting is such a vital stage in the insight process, that a crude sketch on lined writing paper is far better than no plot at all. Remember there is no excuse for failing to plot and look.

7. Don't be afraid to use your own variations on the basic plots. Different topics and different audiences require varying treatments.

As an example of the imaginative uses of box-plots, consider the item from *Which?* shown in Fig. 7.4.

TASK 7.3 Discussion. *Look at the example taken from* Which? *magazine. The aim was to show the effect on life-span of various health hazards. Is it a box-plot? Is it well done? Could you convey the information more clearly, or with more impact?*

7.3 FURTHER VARIATIONS ON THE BOX-PLOT

Dealing with 'far-out' values

In many cases there are one or two 'way-out' values which stretch the whisker far beyond the usual values. One way of dealing with these is to stop the whisker at the normal limits, and add in the 'far-out' values as single points. These points should be individually labelled. Tukey gives an example of the heights of volcanoes which is shown in Fig. 7.5.

Fig 7.4   *report on diet and cancer, Which?, July 1983*

What causes cancer?
The range of estimates shown for the effects of diet is vast, but even at 10%
it would be the second or third biggest preventable cause of cancer.

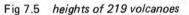

Factors	0	10	20	30	40	50	60	70
Tobacco								
Alcohol								
Diet								
Food additives								
Reproductive and sexual behaviour								
Occupation								
Pollution								
Industrial products								
Medicines and medical procedure								
Geophysical factors								
Infection		Top of range not estimated						

Key

Range of acceptable estimates   Figures in %

Best
estimate

Fig 7.5   *heights of 219 volcanoes*

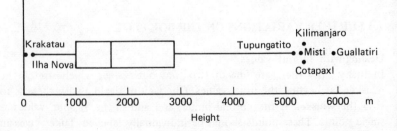

How do you identify what is far-out? Although Tukey gives a formula to calculate these values, it is really a matter of personal judgement and convenience. Very often it is the far-out values which are your main focus of interest, so you may decide to include quite a number of them.

TASK 7.4   Redraw the box-plot for house prices, with a suitable number of 'far-out' values (probably on the high side).

Add upper and lower decile points
A further modification to the standard box-plot is to include the decile points. Just as the quartiles were the quarter points, the deciles are the one-tenth points:

$$\text{Lower decile } D_l \qquad (N + 1)/10$$
$$\text{Upper decile } D_u \qquad 9*(N + 1)/10$$

As before, if this fails to give a whole number, take the average of the values either side. The plot with deciles looks like this, with the whiskers changing to broken lines beyond the deciles:

Examination scores: box-plot:

Rank: 1   2   3   4   5   6   7   8   9   10 11   (bottom up)
      23, 39, 40, 45, 51, 55, 61, 64, 65, 72, 78
      11 10   9   8   7   6   5   4   3   2   1   (top down)
      $N = 11$ $D_l = (11 + 1)/10 = 1.2$; $(23 + 39)/2 = 31$
                 $D_u = 9*(11 + 1)/10 = 10.8$; $(72 + 78)/2 = 75$

Examination scores (showing deciles):

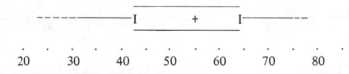

TASK 7.5   Redraw the box-plot for house prices showing deciles (but not 'far-out' values).

7.4 POLYGONS AND CUMULATIVE FREQUENCY POLYGONS

This section is included so that you will understand these terms if you come across them. They have been used for many years, but will probably be replaced by box-plots.

The *frequency polygon* is similar to a histogram. The solid vertical bars you get on the histogram are replaced by points at the top of the bars. These points are then joined up to produce the polygon (polygon

- a figure with many straight sides). Returning to the examination scores example, you will recall we had the following frequency distribution table:

x	f	
		f – the frequency
20–	1	
30–	1	
40–	2	
50–	2	
60–	3	
70–	1	
	11	

The *frequency polygon* plot is shown in Fig. 7.6.

Fig 7.6 *frequency polygon*

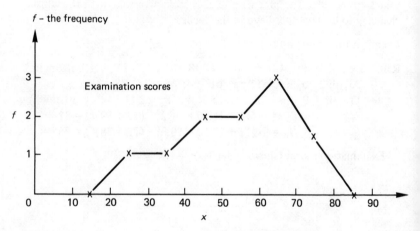

The *cumulative frequency polygon* is an extension of the simple frequency polygon. Go back to the frequency table for the exam. results yet again, and this time add on a *cumulative* frequency column:

x	f	cumf	
			cumf – cumulative frequency
			(like count-up)
20–	1	1	
30–	1	2	
40–	2	4	
50–	2	6	
60–	3	9	
70–	2	11	

11

The graph in Fig. 7.7 shows the cumulative frequency polygon for this data. Note that the scale and plotting points have changed. This is because what the data are saying is different. On the table the entry

$$20- \quad 1$$

now means 'up to a score of 29.99 we have found 1 value'. Similarly, the other intervals have this 'up to' meaning.

Fig 7.7 *cumulative frequency polygon*

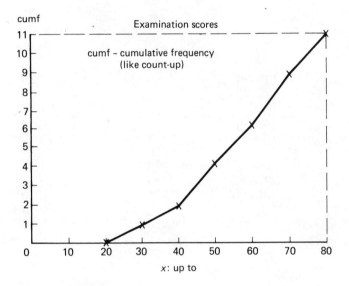

PART III
SUM UP AND
EXPLAIN THE DATA

In this part you will learn about calculations which summarise the data
- explaining them in more 'scientific' terms.

CHAPTER 8

MIDDLE VALUES

8.1 CENTRAL MEASURES

To develop the exploration of your data set, we are now going to analyse the numbers in three ways – to look for middle values, to see how the numbers are spread out, and to look at the shape, or symmetry in the distribution. The topic for this chapter is middle values, or *central measures* to give it its proper name. We are looking for a single value, in the middle, which represents the data set. Because I am going to get you to look at central measures in several different ways, the diagram shown in Fig. 8.1 may help in mapping things out for this chapter.

All the topics shown in Fig. 8.1 are interrelated; you may find it more profitable to skim through this chapter first, and then come back for a more solid attack. In this way you can build up your understanding of central measures, gradually, like coats of paint. In this chapter you need to explore, to reflect, to expand your ideas.

8.2 ISN'T CENTRAL MEASURE JUST THE AVERAGE?

'Average' – add up all the values and divide by the number of values – is *one* form of central measure. There are others, as you will see in the next section. The name 'average' has suffered from careless use, so statisticians avoid using it.

8.3 CENTRAL MEASURES: NAMES

1. Median: M_d – the halfway value
2. Tri-mean: $(Q_l + 2*M_d + Q_u)/4$
   an alternative to the median

3. Mode:
   the most likely value
   the value found most frequently
4. (Arithmetic) Mean: (sum of the values)/(no. of values)
   sometimes called 'average', but not here!

This list is by way of introduction only. In the following sections you will work with these four measures, developing an understanding of each.

Fig 8.1

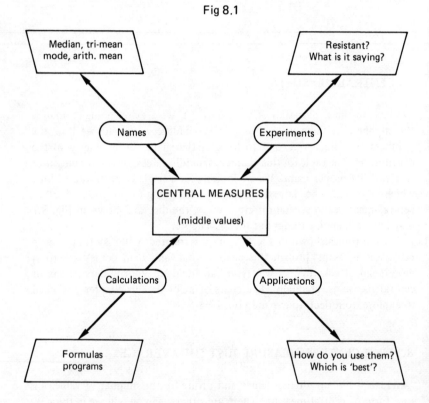

8.4 CALCULATION METHODS

You have worked out the *median* value for your house prices already. The values you need for calculating the *tri-mean* are available too.

For the *mode*, go back to the stem and leaf plot. The stem with the most leaves has the value of the mode (or 'modal' value). There are two problems with this definition:

(1) Either the stem is a wide interval like '£30 000–£35 000' or if it is on a discrete scale gives an unhelpful value like '3 bedrooms'. We could interpolate a more 'precise' value than this, just as we did with the median. But the mode is not all that important, so leave the mode as suggested above, in wide-interval form. (Full details of mode calculations can be found in standard textbooks, for example: Spiegel, *Theory and Worked Examples in Statistics*).

(2) The intervals on the stem and leaf plot can be varied. Choosing different stem steps shifts the value of the mode. Later in Task 8.2 you will investigate this.

TASK 8.1   Calculate the four central measures for both house prices and numbers of bedrooms.

On the computer:

Median – found in Task 6.3

Tri-mean – from the 5-point summaries

Mode – from the stem and leaf plot, wide interval

Mean – requires a program, see below.

The *mean* (or arithmetic mean to give it its full name), is calculated just like the 'average':

$$\text{mean} = (\text{add up the values})/(\text{no. of values})$$

In algebraic terms this can be written as:

$$\bar{x} = \frac{\Sigma x}{N}$$

where $\bar{x}$ – 'x bar' the symbol for the mean, Σ is the 'sum of' – add up all values – so that Σx – means add up all values of x, and N – count up how many values.

Here is a simple program to calculate $\bar{x}$:

```
10 S = 0: N = 0
20 INPUT X
25 IF X<0 THEN STOP
30 S = S + X                    s = Σx
40 N = N + 1
50 PRINT "XBAR = ";S/N;"AFTER ";N;" VALUES"
60 GO TO 20
```

(Of course, using INPUT is not a sensible way of entering your data, when you already have it on DATA lines from way back in Task 2.1.)

8.5 RESISTANCY

A measure is said to be *resistant* if it does not change much when the data values change a little. This sort of stability is not just fussiness on the part of statisticians – in practical terms, a non-resistant measure cannot be trusted.

TASK 8.2   Experiment on the 'resistance' of the four measures.
Mode: *Change the stem interval in the program, and note the value of the mode after each change.*
Median, tri-mean and mean: *Experiment by changing the data set, and using programs to calculate the measures after each change.*

The following changes are suggested, but you should try others:

Add a value just above the median M_d.
Add a value a long way above M_d.
Add two values, one above M_d, one below.
Add two values both just above M_d.
Add two values both well above M_d.
Repeat all these but with the added on value *below M_d*.

Record the results of each experiment and *comment* on the amount of 'resistance' shown by each measure.

Recording could be done on a line scale:

$$\ldots\ldots\ldots 30 \ldots\ldots\ldots 35 \ldots\ldots\ldots 40 \ldots\ldots\ldots$$

$$M_d$$

Show the effect of the changes in the data set by marking the new value of the measure on the scale.

Actually, that task was mainly intended to give you a feel for the measures, and how they are calculated, as well as look at resistance.

8.6 WHAT ARE THESE CENTRAL MEASURES MEASURING?

The mode is the most frequently occurring value. As such, it is the most likely, the most popular, the most common.

The median, as you saw, is the value in the middle when all the data values have been sorted into order.

The tri-mean is just a more resistant version of the median.

The arithmetic mean finds a balancing point in the data. This is illustrated in Fig. 8.2, where there are four houses, one 2-bedroomed house, three 3-bedroomed houses and the aim is to find the mean number of bedrooms.

'The Median, as you saw, is the value in the middle when all the data values have been sorted into order.'

Fig 8.2 *balance point: the arithmetic mean*

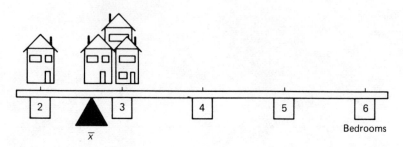

The effect of changing the data set is shown in Fig. 8.3. If we add a 6-bedroomed house to the set the mean moves to the right by a significant amount.

Fig 8.3 *shifting the balance: effect of adding a six-bedroomed house*

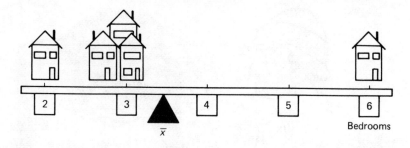

Although the balance point can shift dramatically, the mean is still valid when used in calculation like this:

Area A: Mean number of bedrooms per house = 2.75
        Total number of houses 100

From this information we can work out that there is a total of 275 bedrooms (2.75 × 100) in Area A.

Area B: Mean number of bedrooms per house = 3.40
        Number of houses = 80

Which area has the most bedrooms, A or B?

Only the mean can be used in further calculations; none of the others can be used in this way.

TASK 8.3 *Use the results of your calculations of central measures to answer the following questions concerned with your data set. Write out your answer with reasons in each case.*

*A   Intergalactic Computers Inc. is thinking of setting up a factory in
your area. Their decision will be influence by the cost of housing. They
wish to know the price of an 'average' house.*

*B   Home-owning is a major asset for most owners. In your area, what
is the most likely value/price of a house?*

*C   Tax collectors might need to know the total value of all the houses
in your area (e.g. to gauge the yield of a property tax). If your set of 100
can be assumed to be representative of all, say, 5000 houses in your area,
what is their total value?*

*D   A speculative builder hopes to erect a small number of identical
houses. What do you suggest is the most popular number of bedrooms
he ought to include in each house?*

*E   Overcrowding is said to exist where there are more than 1.5 persons
per bedroom. What is the maximum number of people who could legally
be housed in your 100 properties?*

8.7 APPLICATIONS OF CENTRAL MEASURES

The best way of looking at this is by an example:

W. Pane is employed by the Council to replace broken windows. From
the records of the last few months, his statistically-minded daughter
N. Pane has done some calculations. She reckons the 'average' time to
replace a window-pane (excluding travel time) can be given as:

$$\text{Mean: } \bar{x} = 2.0 \text{ hours}$$
$$\text{Median: } 1.8 \text{ hours}$$
$$\text{Mode: } 1.7 \text{ hours}$$

How can Mr Pane use these figures?

(1) If vandals had broken all 100 windows in a school, it would take him
a total of $100 \times 2.0 = 200$ hours. The *mean* is the measure to use for
this sort of calculation.

(2) If just 1 window is to be fixed, then the most likely time it will take
is 1.7 hours – the *mode*.

(3) Mending windows on a one-off basis is Mr Pane's normal job. If he is
asked to state how long he takes to fix an 'average' window, the
best value to give is 1.8 hours, the *median*. If he uses this value, then
half the jobs takes longer, half take less time.

*TASK 8.4   Consider the negotiations for a wage claim by a large group of
lorry drivers. The term 'average pay' will be bandied about; taking the
mean, the mode and the median values for pay, describe how they could
be used by management, unions and neutral observers to represent the
drivers' pay.*

CHAPTER 9

SPREAD VALUES

9.1 A MEASURE OF SPREAD

Finding a central measure for your data set is not the end of the story; you also need to gauge whether the values are mostly bunched together, or are widely dispersed. For this you need a measure of spread (also known as a measure of dispersion). To help you, the approach to the topic is mapped out by Fig. 9.1.

Once more, the approach to these measures requires you to move about the topics on the branches of the diagram in Fig. 9.1, exploring the meaning and use of measures of spread. In this exploration, your microcomputer is going to be an invaluable ally.

9.2 MEASURE OF SPREAD: NAMES

(1) Range: R – difference between the upper and lower extremes U and L.
(2) Inter-quartile range: IQR – difference between the quartiles Q_u and Q_l.
(3) Mean deviation: – average distance of the values from the mean.
(4) Standard deviation: s – average of the square of the distances from the mean.

If you are a bit baffled by 3 and 4, don't despair – their meaning will become clearer as you complete the tasks.

9.3 MEASURE OF SPREAD: CALCULATIONS

The *range* and *inter-quartile range* can be calculated directly from the 5-point summary, with the aid of a pocket calculator.

For the mean deviation and standard deviation, an algebraic formula is needed, as a first stage in writing a program.

Fig 9.1

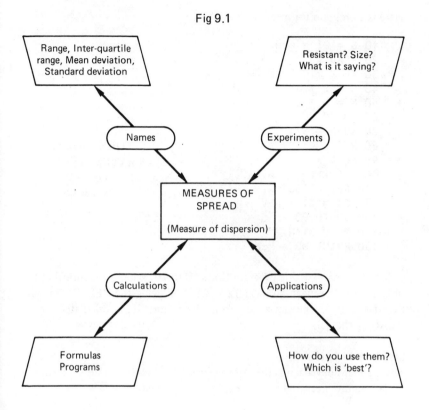

Mean deviation (MD)

The mean deviation is the average (arithmetic mean) of the deviations of the values from the mean. It can be expressed as:

$$\text{Mean deviation} = \frac{\Sigma \, |X - \overline{X}|}{N}$$

This is an 'average' in the same way that $\overline{x}$ was an average, but the top half of the formula needs a bit of explaining:

$X - \overline{X}$ — the distance (deviation) of the values as measured from $\overline{X}$.

$|X - \overline{X}|$ — | | indicates 'absolute value', i.e. taking the positive only, measuring distance, not direction.

$\Sigma |X - \overline{X}|$ — sum of the absolute deviations of all the values in the set.

N — number of values in the set.

Here is a program for mean deviation:

```
10 S = 0: N = 0
20 READ X
30 IF X > 0 THEN 70
40 S = S + X
50 N = N + 1
60 GO TO 20
70 XBAR = S/N
80 RESTORE: S1 = 0
90 READ X
100 IF X> 0 THEN 140
110 DEV = ABS( X - XBAR )
120 S1 = S1 + DEV
130 GO TO 90
140 MD = S1/N
150 PRINT"MD = ";MD
```

If RESTORE is new to you, check it out with your manual - it enables you to read your data again.

The first part of this program (lines 10 to 70) is almost identical to the program for the mean given in Task 8.1. Following the RESTORE in line 80, the sum of the deviations from the mean, S1, is calculated by re-READing the data.

Standard deviation
The standard deviation is the average (mean) of the deviations squared; the deviations are measured from the mean. In formula terms it is:

$$s = \sqrt{\frac{\Sigma (X - \bar{X})^2}{N}}$$

This differs from the mean deviation in *two* ways:

$(X - \bar{X})^2$ - deviations squared (for mean deviation it was absolute deviations)

$\sqrt{}$ - square root needed to preserve the original units, following the squaring of the deviations.

The program for the standard deviation is identical to that for mean deviation, except for two line changes:

```
110 DEV = ( X - XBAR )^ 2
140 SD = SQR( S1/N )
```

Variance: the value of S1/N before taking the square root is sometimes used. This is called the 'variance', and can be found on some calculators on a button marked VAR.

TASK 9.1   Calculate the four measures of spread for house prices and number of bedrooms in your data set. Write programs for mean deviation and standard deviation.

9.4 MEASURE OF SPREAD: EXPERIMENTS

How big? When you were dealing with central measures, you had a fair idea of the *size* of the median, mode or the mean. A particular problem with measures of spread is a lack of 'feel' for their size. This is not made any easier by the fact that the different measures of spread on a set of data give results of widely differing magnitudes. But why don't you investigate the size of these measures of spread for yourself?

TASK 9.2   Investigate the magnitude (size) and resistancy of the four measures of spread, i.e. range, inter-quartile range, mean deviation and standard deviation. Use your data for houses, and experiment on prices and numbers of bedrooms.

Select *samples of 5, 10 and 30, from your data set, using the random number generator (see Task 2.4).*
Calculate *the four measures for each of the 3 samples (Task 9.1)*

Record *the results, preferably as a table like this:*

N	R	IQR	MDEV	SDEV
5	_____	_____	_____	_____
10	_____	_____	_____	_____
30	_____	_____	_____	_____
100	_____	_____	_____	_____

Explore *by taking repeated samples, e.g. 5 samples of 5, 5 samples of 10, 5 of 30 – 15 samples in all. Record the results on a stem and leaf plot.*
Comment *on the size and resistancy of each measure compared with the other measures.*

Before continuing with any follow-up comment, please make sure that you have completed the task – it is crucial in developing your ideas about measures of spread.

9.5 RESISTANCY OF THE MEASURES?

Of the four measures, the range is the most 'natural', and easiest to understand. But is it resistant? Not very, as I expect you found. Just *one* freak value is enough to send the range shooting away. The other three measures are all much more resistant, with not a lot to choose between them. Which was your most resistant measure? The standard deviation usually comes out best.

9.6 SIZE OF THE MEASURE?

As to size, no doubt you found that the range was normally the largest value, with the others in proportions something like this, compared with the standard deviation:

SDEV	MDEV	IQR	Range
1	0.75	1.5	5

Don't be disappointed if your proportions are a bit different; after all this was an *experiment,* which will not always give the same results.

The effect of N, the size of the sample, on the value of the four measures of spread: this is a more subtle distinction, which you may not have noticed. Your sample sizes were 5, 10 and 30, plus 100 taking all the values. As the sample size increased, did you find more consistency? Greater consistency is what you would expect to get, as more information (bigger sample) is available.

9.7 SIZE OF STANDARD DEVIATION FOR VARYING N

But there is one other result of varying the size of the sample – it changes the size of the measure.

TASK 9.3 Investigate how the standard deviation varies with size of sample N, by completing this table.

N	Median SDEV found	Ratio*	Expected ratio[+]
5		$\rightarrow$	1.2
10		$\rightarrow$	1.1
30		$\rightarrow$	1.03
100		$\rightarrow$ 1	1.0

*Ratio $= \dfrac{\text{SDEV found}}{\text{SDEV when } N = 100}$

[+]Expected ratios are values found by theoretical statistics.

As N increases, the value of standard deviation decreases. For N greater than 30 there is little change. This is quite a small variation, which may have eluded your experimental results, but can be established by further experimentation.

Given this characteristic, that the measure depends on the size of the sample, you can either: (a) ignore it – not a bad idea, since the effect is small or (b) adjust the answer to make it consistent.

The 'adjust' option can be found on calculating machines with statistical functions; they usually give you two versions of standard deviation – the 'n' and the '$n - 1$'. For the second one the formula for standard deviation is:

$$s = \sqrt{\frac{\Sigma(X - \overline{X})^2}{N - 1}}$$

Instead of dividing by N, as in the original formula, here we divide by $N - 1$. This gives a calculation of the standard deviation which compensates for the effect of small sample size. For our present purposes though, we are going to stick with the N version.

9.8 WHY $(-\overline{X})$ FOR MEAN DEVIATION AND STANDARD DEVIATION?

Whereas the range and IQR measure the distance *between* two points, for mean deviation and standard deviation the distances are measured *from* $\overline{x}$. In Fig. 9.2 the mean deviation is shown graphically – you measure the

Fig 9.2

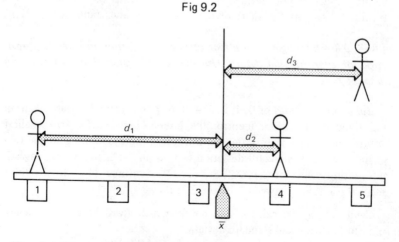

Mean deviation: average (mean) length of deviations
(think of d_1 d_2 d_3 as solid bars)

distance from the datum. For the standard deviation the situation is shown in Fig. 9.3. Here you measure distance again, but calculate distance squared. It is the average of these square areas that gives the value for standard deviation.

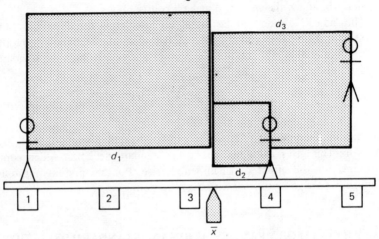

Fig 9.3

Standard deviation: average (mean) of deviations squared
(think of $d_1{}^2$ $d_2{}^2$ $d_3{}^2$ as square plates)

The arithmetic mean is a 'datum' or fixed point for the measurement of deviations. But why choose $\overline{x}$? It seems a fairly obvious choice, but what would happen if you decided to choose some other datum? What effect would it have on the value of mean deviation and standard deviation?

TASK 9.4 Investigate the effect on mean deviation and standard deviation of changing the datum. Use the number of bedrooms from your data set.

Take datum values of 0, 1, 2, 3, 4, 5, 6, 7 and calculate mean deviation and standard deviation. Identify the interval between the two smallest results found, and investigate this interval in more detail, e.g. if the two smallest values of standard deviation were found at 2 and 3, then investigate datum values of 2.1, 2.2, 2.3 ... 2.9. Continue down to the third decimal place, e.g. between 2.78 and 2.79.

Comment on the value of the datum which gives the smallest values of mean deviation and standard deviation.

For this experiment, the program might look like this:

```
10 INPUT "DATUM FOR DEVIANCE"; DAT
20 S1 = 0: S2 = 0
100 FOR I = 1 TO 100
110    READ X
120    DEV = X - DAT
130    S1 = S1 + ABS( DEV )
140    S2 = S2 + DEV. ^2
150 NEXT I
200 PRINT "MEAN DEV ="; S1/100
210 PRINT "STANDARD DEVIATION =";SQR( S2/100)
```

From this experiment, I trust you discovered that the mean, $\bar{x}$, is the datum value which gives the smallest value for the SD and MD. This conclusion could also have been derived using mathematics, but this computerised experiment gives you a nice demonstration. Perhaps you can now also appreciate whey the standard deviation is termed a 'least-sum-of-squares' formula.

9.9 MEASURES OF SPREAD: APPLICATIONS

The *range* is easy to understand and explain, but is very unstable. It is often used where simplicity is the main consideration.

The *standard deviation* is the most complex measure, yet it is the one which is most often used. But what is it measuring? Why take the square of the deviations? Many of the people who use the standard deviation would find it difficult to answer these questions. What can account for the popularity of this measure?

(1) It is resistant.
(2) Calculation is not difficult if you have a statistical function calculator.
(3) Standard deviation becomes very important when you move on from *exploring* the data, as we are doing in this book, to drawing scientifically valid conclusions (*inferences*).

The *IQR* is a good compromise. It is much more resistant than the range. It covers the middle 50 per cent of the data. 'Half the values (i.e. between Q_u and Q_1), is a readily understood statement.

The *mean deviation* is seldom used. Actually, I only slipped it in because it is a useful stage in understanding the standard deviation. It is also useful for you to realise that there are other measures besides the ones in general use.

TASK 9.5   Reconsider the window-fixing activity of Mr W. Pane in the last chapter. How might he use the measures of spread to compare the time

*per window in the vandalised school, with the regular one-off jobs he
also receives?*

Measures of spread are just that – measures. As such they only come
alive when comparing things. Compare this with the way the power output of cars is measured. If I tell you that my car has a 50 kW (kilowatt)
engine, this may not tell you much. But if I tell you that a high-powered
sports car has a 200 kW engine, you have a basis for comparison. In the
same way measures of spread are values without a great deal of interest
on their own, but useful when making comparisons.

CHAPTER 10

SKEWNESS AND RE-EXPRESSION

10.1 SHAPE OF THE DISTRIBUTION ON THE STEM AND LEAF PLOT

In this, the last chapter in the part on summarising the data, we are going to look at one further topic - skewness. As well as discovering methods of spotting skewness, you will also encounter a technique called 're-expression' which is useful when dealing with skewed data, and in other circumstances too.

10.2 SKEWNESS: NAMES

When looking at the shape of the distribution of the data that might be found on a stem and leaf plot, you could recognise:

(1) *Symmetry*   When the values are spread evenly above and below a central value, say the median or mean.
(2) *Skewness*   Found when there is an uneven 'straggle' - a long tail on one side of the distribution, with values heaped up on the other side.

Skewness can occur in two ways:

   2.1 *Positive skew*: straggle in the high values.
   2.2 *Negative skew*: straggle in the low values.

10.3 IDENTIFYING SKEWNESS

We will be looking at methods of measuring the extent of skewness. But first it is more important to be able to identify skewness, to recognise when a data set is skewed. Our old friends the boxplot, and the stem and leaf plot are especially useful.

Examination results:

Group 1		Group 2

10s	Units
1	1
2	46
3	1
4	13776
5	0258
6	59
7	
8	

```
      ——I   +    I————————

     10   20   30   40   50   60
```

Lowside straggle   Highside straggle

Neither of these distributions of exam. marks is symmetrical; they are both skewed, but in varying ways. There are degrees of skewness – highly skewed, slightly skewed, not skewed (symmetrical).

TASK 10.1   From the plots (box and whisker or stem and leaf) of the distribution of prices, number of bedrooms and garage facilities, judge whether the data in your set are skewed.

10.4 SKEWNESS AND CENTRAL VALUES

Knowing whether your distribution is skewed or symmetrical is interesting, but much more significant is to appreciate the effect skewness has on the central measures. If you have a nearly symmetrical distribution, then the median, the mode and the mean will have almost exactly the same value; as the distribution becomes more skewed, the three values spread out. To verify this, carry out the following experiment.

TASK 10.2   Take a fairly symmetrical distribution (I would expect the number of bedrooms to be like this; if they are asymmetrical (not symmetrical), cheat a little and trim off the straggle).

Calculate and record the mode, median and mean for the symmetrical distribution.

Skew the distribution deliberately by replacing values near the middle with straggly 'tail' values. Do this first on the upwards side with just a few (5 – 10) values, then quite a number (20 – 30), to produce marked skew. Returning to the original symmetrical data set, repeat the process on the downwards side.

Record *the effect of these experiments on the central values:*

Experiment	Mode	Median	Mean
Symmetric (original)			
Slight up skew			
Serious up skew			
Slight down skew			
Serious down skew			

Comment *on the order of the measures after the deliberate skewing (a graphical display might help)*

This effect of skewness in spreading out the central measures is very important. In the real world you will find that more often than not your distribution in skewed. If you find that the mode has a smaller value than the median, with the mean having the highest value, this indicates highside straggle. (Is this what you found with your experiment? For lowside straggle the order is reversed.) You can see now what problems can arise when people start talking about 'average wages'. Wage and salary distributions nearly always show highside straggle – so average in the sense of mean will be the highest possible measure – this makes it a big favourite with the management. 'But more of my members earn £x' replies the Union Negotiator, quoting the mode. In these circumstances, the median is the fairest, most truly representative value of the wage of the person in the middle.

10.5 DEALING WITH SKEWNESS: RE-EXPRESSION

One way of dealing with skewness is to iron it out – to reduce the distribution to the symmetrical shape. This can usually be achieved by a process of transformation, by 're-expressing' the data in another scale. This involves mathematical transformations, which may sound a bit tricky. Although the theory requires some advanced maths, on the computer the technique of re-expression is very simple. Make a few experiments with your data on the computer and you will soon get the hang of it.

The mathematics of re-expression works like this: READ a value, then transform it using a suitable function. The square root is a very popular transformation. So if you start with an evenly spaced scale and take the square root of the values, you finish up with a shortened scale, with the high values pushed together as shown in Fig. 10.1.

TASK 10.3 *Investigate the effect of a square root re-expression on the house prices in your data set.*

Fig 10.1

Raw (original) scale

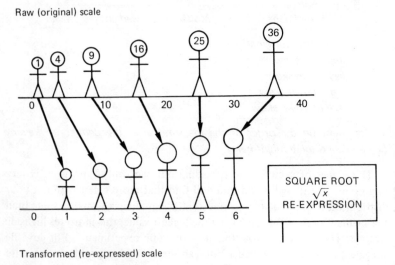

SQUARE ROOT
$\sqrt{x}$
RE-EXPRESSION

Transformed (re-expressed) scale

Comment *on the effect of the re-expression on the skewness, by examining the plot before and after re-expression.*

For this task you could make use of the ordered set of data you worked out in Tasks 5.4 and 5.5. Work out the stem and leaf plot as before. Then Re-READ the data, but this time transform the data like this:

```
10 READ X
20 X = SQR( X )
```

Note that the scale and steps will be much smaller after the re-expression, so remember to make allowances for this.

There is no way I can know whether your house price data was skewed or symmetrical, but I would guess it was skewed, straggling on the upwards side.

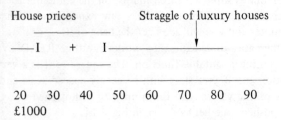

Square roots are good at 'curing' this sort of upward straggle. On the other hand, if your data showed symmetry or downwards straggle, then the transformation would only have made things worse!

10.6 **STAIRCASE OF RE-EXPRESSIONS**

There are many other ways of re-expressing the values – after all you have a whole page of 'Algebraic Functions' in your programming manual. Which re-expressions are useful? In textbooks on exploratory data analysis (e.g. Erickson and Nosanchuk, *Understanding Data*), you will find the staircase, or ladder of re-expression transformations as shown in Fig. 10.2.

Fig 10.2

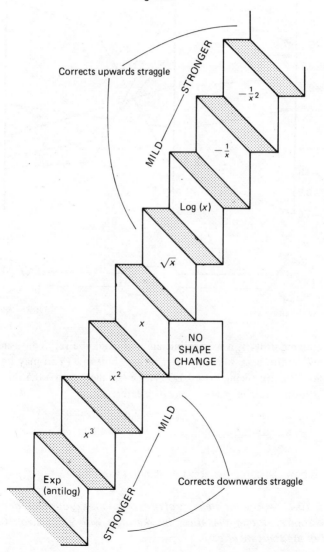

The idea is that you can go up or down the ladder, re-expressing your data until a satisfactory shape is found – just how much effect each step has will be shown by the next experiment.

The reason why the '1 over' re-expressions are written with a minus sign is to keep the order of the data. Fig. 10.3 shows what happens with and without the minus.

Fig 10.3

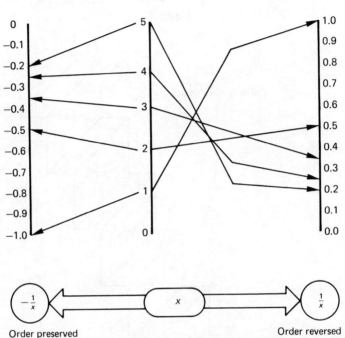

Order preserved   Order reversed

Preserving order is a worthwhile aim – it would be very confusing if the top end of the data suddenly became the bottom. (You may be having a little difficulty seeing how –1.0 can be 'smaller' than –0.5, but if you look at it on a number scale it may be clearer:

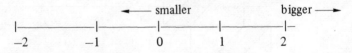

(This is how the computer sees the numbers as well.)

TASK 10.4   Apply the staircase of re-expressions to your data for numbers of bedrooms, noting the effect on skewness with each 'step'. Can you confirm the staircase effect?

Does the idea of transforming a number of bedrooms bother you? Perhaps the thought of $\sqrt{3}$ bedrooms strikes you as odd, if not impossible. But remember Chapter 4 on scales of measurement. All scales are arbitrary to some extent; square root or any other re-expression does not destroy the essential nature of your data, and is often a great help in exploring it.

10.7 APPLICATION OF RE-EXPRESSION

You have seen the main application of re-expression already – in exploring the amount of skewness in your data. Re-expression is also a useful tool to keep at hand when the data scales become inconvenient. For example, if most of your house prices were bunched in the £30 000–40 000 interval, re-expressing using the X^2 would stretch out these values. (Of course, you could achieve the same effect by simply changing the interval size on the computer stem and leaf plot.)

PART IV
COMPARING DATA

In this part we move on from your single set of data, and look at ways of comparing sets of data.

CHAPTER 11

COMPARING

DISTRIBUTIONS

So far we have been exploring and analysing a single set of data on its own. Data almost never exist on their own; there is generally some background, some other data for comparison. Even within a single set of data, it is possible to extract subsets which are natural groupings. Comparisons can then be made amongst these subsets. This chapter will describe methods for comparing sets of data. But in order to have some data to work on, you need to collect some more information on houses.

TASK 11.1   Collect a random* sample of about 25 houses-for-sale, from the same area as your original 100 houses. This new, smaller sample will be separated from the original sample in time – that is the basis for comparison.

(I'm assuming that some time has passed since you took your original set of data. A month or so would be the minimum gap as far as this task is concerned. If you have reached this point sooner than that, try to find *earlier* data – from an old newspaper, for example.)

Record the same information as in Task 1.3, i.e. price, number of bedrooms, garage facilities. (Please note that this data collection exercise will be repeated in the next chapter, so be prepared!)

*Random means that each house advertised for sale in your area has an equal chance of being selected. So, if there was a total of 250 advertisements, picking out every 10th one should yield a random sample of 25. If you really want to be clever, you would number all the advertisements (1 – 250 say), and select the sample using the Random Number Generator on your computer:

```
10 FOR I = 1 TO 25
20 PRINT INT( RND(0) * 250 + 1)
30 NEXT I
```

So long as your sample is fairly random it will be all right for our present purposes. But be warned! Not choosing a truly random sample has been the downfall of many a survey, including some elaborate and expensive surveys carried out by highly reputable investigators. If you want to know more about methods of conducting a survey and choosing a sample, read through Chapter 16, 'Mistakes, Errors and Surveys'.

11.1 COMPARING NEW WITH OLD

To compare the new set of data with the original set, we can make more use of the stem and leaf plot.

Example: Exam. results:

English: 23, 39, 40, 45, 51, 55, 61, 64, 65, 72, 78
Maths: 37, 41, 44, 48, 58, 61, 63, 69, 75, 83, 89

How did the two groups get on? One way we can see is by side-by-side stem plots:

English:			French:	
Stem (10s	Leaf (1s)		Stem (10s)	Leaf
2	3		2	7
3	9		3	7
4	0 5		4	1 4 8
5	1 5		5	8
6	1 4 5		6	1 3 9
7	2 8		7	5
8			8	3 9
9			9	

The comparison can be made more dramatic by back-to-back stem-and-leaf plotting:

English		French
3	2	
9	3	7
5 0	4	1 4 8
5 1	5	8
5 4 1	6	1 3 9
8 2	7	5
	8	3 9
	9	

(stem 10s leaf 1s)

TASK 11.2 Plot the original house prices and the latest house prices as a pair of back-to-back stem-and-leaf plots. Use the plot to judge whether the general level of prices has changed.

'To compare data we can make more use of the stem and leaf plot.'

84

One difficulty you may have had with this task is the difference in size between the new and original sets of data. What you need to do is to look at the shape of the two distributions, and *imagine* how they would compare if they both contained the same number of items.

11.2 MULTIPLE BOX-PLOTS

Back-to-back stem plots are limited to comparing *two* sets of data. Box and whisker plots, the other main tool for looking at data, are much more adaptable. One example of a kind of multiple box-plot comparison was given in Chapter 7 in the example taken from *Which?* ('What causes cancer?'). Below is another example:

Salaries of statisticians: Results of a survey on 100 members from each of the National Statistics Associations.

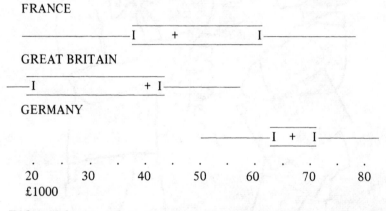

FRANCE

GREAT BRITAIN

GERMANY

20   30   40   50   60   70   80
£1000

TASK 11.3 Investigate the effect of garage facilities on house prices, ie the distribution of prices for houses with no garage, single garage and double garage. Use multiple box-plots to show the comparisons. Comment on the medians, spreads and skewness found.

For this task you need to extract *3* files of data from your original set, one for each of your classifications of garage, i.e. 'no garage', 'single garage' and 'double garage' coded 0, 1 and 2. (You may have used a slightly different coding scheme.) So READ in the DATA for all the houses, and *if* the garage code is ∅ (the code for 'none'), *then* add this house to the file of no-garage houses. When you have found all such houses, you can draw the box-plot. Repeat for the other two categories, and you will have three box-plots which together will help you decide what effect garages have on house prices. You don't have to be over-clever and get the computer to do all of this plotting; it may be a lot simpler, and more appropriate, to draw a stem and leaf plot by hand, using the values found by the computer, and then draw in the multiple box-plots again by hand. You must decide which is the more suitable technique given your own facilities.

The remainder of this chapter may be omitted, if time is pressing.

11.3 NOTCHED BOX-PLOT

Using the back-to-back stem and leaf plots or the multiple box-plots for comparison is a natural extension of the previous work. Looking at these plots is an essential step in judging whether prices have changed between taking samples, or whether garage facilities have a detectable effect on house prices. You may look at the plots and feel convinced that prices have, say, gone up a little, but can you be sure?

This leads us on to a new area – confirmatory statistics. In this book we are mainly concerned with exploratory statistics, looking at the data in various ways and getting a feel for what they are saying. In confirmatory statistics we attempt to prove that, for example, prices have really changed. Notched box-plots are a small excursion into the realm of confirmatory statistics.

Returning to the question, 'Can you be sure that house prices have gone up?' prompts the reply: 'Why should you be UNsure?' To explain this, look back at the multiple box-plots showing salaries of statisticians. The Germans are undoubtedly paid more than the French or British. But are the British really paid less than the French? There is a lot of overlap on the box-plots, and the median values are not all that far apart. A lot of uncertainty surrounds all such sets of data, so the question you should ask is, 'If we were to take repeated samples, and knowing that sample results are variable, what is a believable range for the median value for the salaries paid?' The answer to this question might be:

British: Median found in sample £45 000
Lowest likely median £41 500
Highest likely median £48 500

This can be illustrated on the box-plot using 'notches' for these likely values:

GREAT BRITAIN:

```
—I                  >     + I–<————————

    .       .       .       .       .       .       .       .       .       .
   20      30      40      50      60      70      80
  £1000
```

The $>$ and the $<$ represent the lowest and highest likely values for the median. How are they calculated? The following experiment will show you.

TASK 11.4   Take a random sample of 10 houses from your original data set (you have to use the computer for this). Calculate the median price for this sample.

Repeat this process for a further 19 samples, giving a total of 20 samples, with a median price calculated and recorded for each. Plot these 20 median values on a stem and leaf plot.

Comment on the variability of the median values found and compare them with the value found in Task 8.1.

This experiment brings out the fact that even when samples are taken from the same set of data, they do not always give the same result. There is always uncertainty about any measurements based on simple data. The best we can hope for is to put some limits on this variability. The convention for these limits, and the values used for the 'notches' on the box-plot are to take the bottom 5 per cent limit, and the top 95 per cent limit. In the last experiment there were 20 samples, and one-twentieth is 5 per cent. So if you locate the 2nd highest and the 2nd lowest values for the median that you found, this will give the notch values.

11.4 FORMULA FOR NOTCHED BOX-PLOT

Doing this kind of experiment to gauge the likely variability of the median is both cumbersome and slow (although not too difficult on the computer). There is a formula method for working out the notches which is:

Notch at: Median $\pm 1.58 \times IQR / \sqrt{N}$

Example: Median at £45 000

$$IQR = Q_u - Q_l = £47\,000 - £25\,000 = £22\,000$$
$$\sqrt{N} = \sqrt{100} = 10$$

Notches at: £45 000 $\pm$ (1.58 $\times$ 22 000 / 10)

= £45 000 $\pm$ £3476

upper notch at £48 476; lower notch at £41 524

TASK 11.5   Calculate the notch values for house prices by formula. Take the IQR as found in Task 9.2, and take the size of sample to be 10. Compare the notch values found by formula with those you found in the experiment in Task 11.4.

Also *for the multiple box-plots of the effect of garage on house price (Task 11.3), calculate, using the formula, the notches for all the box-plots. Do the notches change your view on the effect of garage on price?*

The following comments apply to the last two tasks, both of which were concerned with notched box-plots:

(1) Although the experimental and formula values for notches were not exactly the same, would you accept that they are of the same general size (i.e. can you accept that the formula might work)?
(2) The notch limits are very wide, possibly much wider than the variation you may have expected. Are these limits too cautious? If you want to be really sure that a house with a garage is worth more than one without, do the notches tell you something you couldn't see from the box-plot anyway?
(3) Does the formula $(1.58 \times IQR / \sqrt{N})$ look reasonable? There are three parts to it, so let's look at each in turn:

(a) $/ \sqrt{N}$: divide by the square root of N, the size of the sample. As the sample gets bigger so the notch-spread reduces.
(b) $\times IQR$: multiply by the inter-quartile-range. So if IQR is big the notches are wide, if IQR is small the notches are closer to the median.
(c) 1.58: a constant value of about one and a half. The notch-spread will be of the same sort of size as the IQR.

The formula looks as if it could make sense, don't you think? It has been derived algebraically, so it can be 'proved' to be right, but you should be convinced it is valid before you accept it.

CHAPTER 12

INDEX NUMBERS

You can hardly avoid index numbers – scarcely a day goes by without some mention of them on the TV news, or in the daily papers. The movements of the Stock Exchange are reported from London by the Financial Times Share Index, or from New York by the Dow Jones Average. Changes in the cost of living are reported once a month, when the latest value of the Retail Price Index (UK) is published.

In this chapter you are going to use your house price data to explore ways of constructing and calculating index numbers. The layout of this chapter is shown in Fig. 12.1.

12.1 DEFINITION

'An index number is a measure which summarises the changes in value of a set of items.'

You have seen summary measures before – remember such measures as the median or the inter-quartile-range? An index number is also a measure. It measures the change in level of a whole set of items compared with some base period.

'Value' takes us into the realms of economics, where we must distinguish between value-in-exchange – the price of a thing, and value-in-use, the worth of a thing. Many of the problems and misunderstandings in index numbers are caused by the double function that index numbers have – to be both a valid statistical measure, and a valid economic indicator.

Indexes are usually based on a whole range of items. Indeed, one of the main purposes of index numbers is to simplify a large number of changes into one simple, easy to understand single figure. In order to calculate an index, it is usual to take a 'basket' of items – a small selection which is taken to represent all of the items in the set. In this chapter we will be looking at just one item – house prices. Because houses are not all the same, it will be necessary to choose a 'basket' in just the same

Fig 12.1

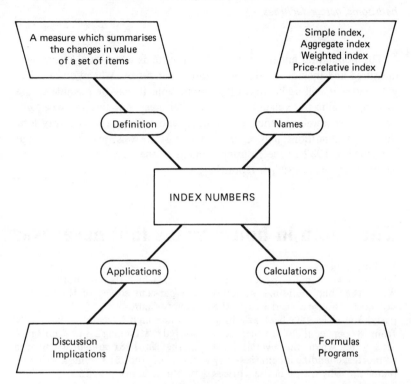

way as you would if you were doing a survey on the level of prices in the shops.

12.2 POPULATION

An index can only measure what it sets out to measure – that seems pretty obvious doesn't it? So what are you looking at with your data base of houses? Back in Task 1.2 you noted the *area* where, and the *time* when you recorded your data. Any index you calculate should be used only in measuring change for that time and place. A further restriction may be that you did not cover all types of housing – perhaps the adverts did not include apartments or flats.

TASK 12.1   (This is a repeat of Task 11.1)
Collect a second random sample of about 25 houses-for-sale from the same area as your original 100 houses. This second small sample will be separated from the first sample and the original sample in TIME – that is the basis for comparison.

Record *the same information as in Task 1.3, i.e. price, number of bedrooms, garage facilities.*

The population under study is all of the items which could have been measured and included within your sample or survey. When you calculate your index it will be based on this population. It may be possible to use this index value to explain changes in other areas, but care must be exercised. Index numbers, once established, tend to be used for a very long time, and conditions may change. To illustrate this, just look at what happened in 1982 to the Housing element of the Retail Price Index (UK) as shown by the newspaper report below.

The boom in house prices that never was

By Tim Roberts

WHAT makes a house price boom? With the big building societies saying that house price rises over the past 12 months have been anything from 8 per cent to 15 per cent at a time when the overall level of mortgage lending is being severely reduced, why are house prices increasing?

Indeed, are they increasing at all? Estate agents and surveyors are reporting a much lower rate of house price increase, if any at all, while the building societies and the Government continue to report a new house price boom.

So who is telling fibs? The answer is: nobody. But the problem really lies in the way that the statistics are collated and in the major role that the banks have played in distorting the mortgage market over the past couple of years.

The building societies' own figures are based on their mortgage lending: specifically, the purchase price of the property. So is the Department of the Environment's own House Price Index, which is no more than a five-per-cent survey of the building society figures.

The high street banks really started attacking the mortgage market in 1981 and provided one third of the £13 billion of new mortgage loans in 1982.

But the banks largely displaced building society mortgage lending in the upper price bracket.

Building society statistics show that in 1982, 87 per cent of all mortgages were on properties under £36,000.

The effects of this has seen to severely depress the average value of houses on which the societies have made loans. That inevitably follows through to the house price statistics.

There is a very strong case to be made that there was no house price slump in 1982 in the face of good supplies of mortgage finance. In fact, probably the reverse is the case.

From the *Guardian*, 18 July 1983

12.3 SIMPLE INDEX

A simple index follows the changes in a single item. Example: Consider Acacia Avenue, a long road of identical houses, with a constant turnover of houses:

Prices

	In original survey	In Task 11.1	In Task 12.1
	p_0	p_1	p_2
	£35 000	£37 000	£34 000
Index	100	$\dfrac{37}{35} \times 100 = 105.7$	$\dfrac{34}{35} \times 100 = 97.1$

TASK 12.2   If suitable data are available, calculate a simple index for a typical property.

This technique of following the price of one typical property is regularly used by valuers and estate agents (they call it 'comparable' property), but it is obviously limited in scope and accuracy.

12.4 AGGREGATE INDEX

It would be better if we took all the data into account and compared the 'average' price change, rather than just the change in one property, however typical. (I have put 'average' in quotes to remind you that we have several central measures – mode, median and mean, for example). 'Aggregate' indicates a collection or group of items.

The aggregate index is conventionally calculated by taking the means of the various samples. The median value would also be acceptable; in many ways more suitable for the job in hand.

The calculations are in two stages: first, calculate the aggregate or average value for the base period, and each of the subsequent periods. Second, calculate the index in exactly the same way as for the simple index, but using the aggregate value instead of the single value.

TASK 12.3   Calculate the aggregate index values for house prices for your two samples of houses, using the original data set as the base. Use both mean and median aggregates.

Comment on which aggregate measure – mean or median – is more appropriate in explaining the changes in price level of a variety of houses?

12.5 **WEIGHTED INDEX**

You may have felt uneasy when calculating the aggregate index by just lumping all the values together. For instance, sample variability, which always occurs, may have prevented you from comparing like with like. Perhaps the second sample contained many more expensive houses compared with the first. This would create a much larger rise in the index than the change in prices that actually took place, even if all houses showed the same percentage rise.

For example, two samples, both containing two items:

Sample 1 £20 000   £80 000   average £50 000 = 100
Sample 2 £60 000   £60 000   average £60 000 = 120

Sample 2 contains two higher priced houses; does a 20 per cent rise in the index indicate anything, or have these results been drawn from different ends of the property market?

To overcome this problem, the technique of weighting is used to give a fair representation to the different sectors or subsets of the market. First you have to decide the basis for identifying the sectors. For the houses in your sample this could be on the basis of numbers of bedrooms. Thus there will be a separate category for houses with 1 bedroom, 2, 3, 4 bedrooms and so on.

In fact, there are probably few houses with just 1 bedroom, so you could combine 1 and 2 bedrooms together. You could also have a catch-all category 4+, for all houses with 4 or more bedrooms. This gives three main sectors:

1 and 2 bedrooms   3 bedrooms   4 or more bedrooms

Next you need to calculate the weights, that is the percentages which correspond to each sector. In fact, you have already done this in Task 3.5. For each sector, you should calculate an aggregate value. So there will be the price of an average 1–2 bedroomed house, the average for 3 bedrooms and average for 4+. Finally, combine these averages with their weights as shown in this example:

	Weighting $w\%$	p_0	p_1	p_2
1/2 beds	20	£20 000	£22 000	£24 000
3 beds	50	£30 000	£30 000	£31 000
4+ beds	30	£40 000	£38 000	£39 000

$$I_1 = \frac{£22\,000 \times 0.20 + £30\,000 \times 0.50 + £38\,000 \times 0.30}{£20\,000 \times 0.20 + £30\,000 \times 0.50 + £40\,000 \times 0.30} \times 100$$

$$= \frac{4\,400 + 15\,000 + 11\,400}{4\,000 + 15\,000 + 12\,000} \times 100$$

$$= \frac{30\,800}{31\,000} \times 100 = 99.4$$

and

$$I_2 = 101.6 \qquad \text{do you agree?}$$

TASK 12.4 Calculate a weighted price index for your house prices. Use your original data set to establish the weights. Calculate an index for both of your follow-up samples using the original set of data as the base.

12.6 WEIGHTED PRICE-RELATIVE INDEX

Although the weighted index produces greater consistency, it still over-represents the changes in price of the more expensive properties. At this point you have to ask, 'Is the index meant to represent the money change in house prices, or should it show the average percentage change?' If it is the money change then a £2 000 rise on an expensive house is equivalent to a £000 rise on a cheap one. The weighted index of the last section will measure this money-value change.

But you might argue that it is the percentage rise that matters. A £2000 rise on a £20 000 house is 10 per cent; on a £40 000 house it is only 5 per cent. This idea leads on to the weighted price-relative index, which gives each sector equal weighting.

The calculations are very similar to the last section; but instead of aggregate sector prices, enter an index value for each sector:

	$w\%$	p_0	p_1	p_2
1/2 beds	20	100	110	120
3 beds	50	100	100	103.3
4+ beds	30	100	95	97.5

$$I_1 = \frac{110 \times 0.20 + 100 \times 0.50 + 95 \times 0.30}{100 \times 0.20 + 100 \times 0.50 + 100 \times 0.30} \times 100$$

$$= \frac{22 + 50 + 28.5}{20 + 50 + 30} \times 100 = 100.5$$

$$I_2 = 104.9 \qquad \text{do you agree?}$$

TASK 12.5 Calculate a price-relative weighted index, following the method used in the last Task (12.4).

12.7 INDEX NUMBERS: SHOULD THEY BE SCRAPPED:

There are more people working on index numbers than all the other parts of statistics put together. Indexes such as the Retail Price Index are of crucial importance politically and financially. To the old-aged pensioner, the saver in 'index-linked' bonds, the index has a real impact. But index numbers are also looked at with contempt by academic statisticians as mere number juggling, tricks with figures which prove nothing. You have seen how indexes depend crucially on choice of population, choice of weighting, and even method of calculation. So are the academics right? Are index numbers so unreliable, so subject to personal judgement that they tell us nothing, that the 'whole bag of tricks ought to be scrapped?' (Moroney in *Facts from Figures*, Penguin, Harmondsworth, 1951)

A change in an index number is not scientific 'proof' of change. It is always possible to alter the basis for the index – the 'basket' – and come up with a different answer. Indexes depend on many value judgements. But an index is of great use because it allows us to explore the data, get an indication of the general level of prices. Judging index numbers by the scientific criteria of confirmatory statistics is to misunderstand their true nature and purpose. Index numbers are an aspect of exploratory statistics. So long as the judgemental element of index numbers is appreciated, they should not be misused. Alas! once politicians or administrators get hold of a number, especially from a computer, they tend to treat it as revealed truth!

Index numbers are used in the real world, and their values cause substantial changes in the fortunes of a large number of people. It is easy to pretend that index numbers represent objective truth, and that any shortcomings can be put right by more measurement. But in this, as in all exploratory statistics you investigate the numbers to gain understanding. Statistics helps you to understand what is going on, but can never be the final answer. It is up to decision-makers such as politicans to choose to index-link savings and pensions; they must also decide which form of the index number is to be calculated.

12.8 APPLICATIONS

In this chapter you have looked at a simple example of an index number which could be used to monitor house prices in your survey district. You would find it rewarding to look in more detail at one of the major indexes in use: the Retail Price Index, or its equivalent for example in the US, the Consumer Price Index. See how it is constructed, and how much effort and cost goes into calculating it. You will find details in books devoted to business statistics, for example, *Modern Business Statistics* by Imam and Conover.

One point you will soon appreciate is how much data are required for these major indexes. It is quite common for details of several thousand prices to be recorded each week. This volume of data can only be handled quickly and cheaply by computer. Of course, once the data are in the computer, further calculations can be carried out at very little cost. This represents a great opportunity – in the future there is no reason why you should not explore all kinds of index numbers, which are tailored to your own requirements. Let the Government Statistical Service provide you with the raw data; you will be able to apply weightings to measure your own changes in living costs!

PART V

RELATIONSHIPS

IN DATA

In this part we see how relationships between sets of data can be explored and analysed.

CHAPTER 13

TABLES

A very popular method of displaying results is by means of a table. You have probably seen this sort of presentation many times:

Examination results for three subjects at three types of school. Number of passes at 'O' level

	French	English	Maths
Public School	763	1205	996
Grammar School	98	205	543
Comprehensive	3431	2641	1901

Although simple in concept, you would have to study this table for a long time before you could make it yield any information. It is quite common to be presented with tables like this, often with many more rows and columns, and no clue as to what information is contained in the table.

The table given above contains just 3 rows and 3 columns. The rows run horizontally, and represent the types of school. The columns are vertical, and represent the subjects. So this table looks at numbers of passes in *two* ways, and we can call this a *two-way* table. It would be easy to find another classification for these results. For example, the passes could be broken down by sex (boy/girl) as well as by schools and subjects (three-way table) Adding information about grades of pass will give a four-way table ... the possibilities are almost limitless. The basic problem will still remain; how do you turn a table of figures into some meaningful *information*, or even into *knowledge* about the subject?

Before tackling this problem, it is time to draw up a table of your own!

TASK 13.1 Using your original data set and results from previous tasks, produce a table to show numbers of houses classified by number of bedrooms and garage facilities. Use pencil and paper.

TASK 13.2   Repeat Task 13.1, but get the computer to draw up and print out the table.

The reason for asking you to do the table, first by hand and then by computer is twofold:

(1) Doing it manually gives you some idea of the programming technique.
(2) You need an independent check on your program, to see that it is working all right.

The program is needed, because later in this chapter the tasks will require repeated operations on the table. But be warned! This sort of programming has to be done very carefully; it is very easy to get the rows and columns mixed up.

The program for Task 13.2 should make use of a two-dimensional array. The 'elements' or 'cells' of this array will become the entries in the table. You could use the array A(I,J), where subscript I represents bedroom numbers (1, 2, 3, 4, 5, 6), and subscript J is garage type (0, 1, 2). But first make sure that the cells are empty.

```
 5  DIM A(7,3)
10 FOR I = Ø TO 7
20   FOR J = Ø TO 3
30     A(I,J) = Ø
40   NEXT J
50 NEXT I
```

Now the counting can begin:

```
100 FOR I = 1 TO 1ØØ
110   READ P, B, G
120   IF B = 1 AND IF G = Ø THEN A(1,Ø) = A(1,Ø) + 1
200 NEXT I
```

would count how many 1-bedroomed, no garage houses are in your set. You *could* add on 5 more statements to test for other bedroom numbers, 6 more for 'single garage' and bedroom number, and 6 more for double garage – that's 18 statements in all – far too clumsy and tedious. Look at this trick:

```
110 READ P, B, G
120 A(B,G) = A(B,G) + 1
```

So if B = 1 and G = 1 then the value in A(1,1) will be increased by 1. The same will apply to other values of B and G. This may be a bit difficult to understand – it may seem strange that you are using a value to identify the address to be altered. But this sort of programming trick is extremely useful, and is well worth learning.

You *could* also find the row and column totals: these can be put in an extra row or column at the end of the table. To count up the total of houses with single garages (row code J = 1):

```
300 FOR I = 1 TO 6: A(7,1) = A(7,1) + A(I,1): NEXT I
```

(The technique of adding on a row or column is called 'augmenting the array'.)

Repeat the count for the rest of the rows, and also for the columns, making use of nested FOR . . . NEXT loops.

The print-out can make use of nested loops as well:

```
200 FOR J = 0 TO 3
210    FOR I = 0 TO 7
220       PRINT A(I,J),
230    NEXT I
240 PRINT " "
250 NEXT J
```

This program segment is designed to produce a table with garage facilities across the top, number of bedrooms down the side. By changing the order of I and J, you can change the layout of the table.

Taking all the parts into account, this is quite a complicated program, with its nested FOR . . . NEXT loops, and subscripts used in an interesting way. When you have completed the program, you *must* check the output. It is very easy to produce results which look correct, but are switched about in unexpected ways.

13.1 TAMING THE CLUTTER: CODED TABLES

The numbers in the tables do not, on their own, tell us very much. You may be able to pick out the unusually high or low values, but why not get the computer to do that? If you take all the values in the table you can treat them as if they came from a single distribution. Using the box-plot, you already have a method of marking some values. To identify the hinge points requires a stem and leaf plot, so looking at the data again:

	French	English	Maths
Public School	763	1205	996
Grammar School	98	205	543
Comprehensive	3431	2641	1901

Stem and leaf		Box-plot
0*	098 205	
·	543 763 996	———I + I———
1*	205	98 374 996 2271 3431
·	901	
2*		
·	641	
3*	431	SYMBOL: — · +
·		

Values in the table can be replaced by symbols as shown:

Between Q_u and Q_l replace by   ·
Above Q_u replace by   +
Below Q_l replace by   —

As the main area of interest is in the 'whiskers', these are further classified as follows:

'Outside high' defined as above $Q_u + 1.5 \times IQR$; symbol ++
$2271 + 1.5 \times (2271 - 374) =$
$2271 + 2845.5 = 5116.5$

'Outside low' defined as below $Q_l - 1.5 \times IQR$; symbol =
$374 - 2845.5 = -2471.5$

'Far out high' defined as above $Q_u + 3 \times IQR$; symbol P
$2271 + (2 \times 2845.5) = 7962$

'Far out low' defined as above $Q_l - 3 \times IQR$; symbol M
$374 - 5691 = 5317$

(Of course, the negative values have no meaning here, but are included to show the calculations.)

This classification is similar to the notched box-plots of Chapter 11. There are seven symbols in all:

$$M = - \cdot + ++ P$$

= is double —, M is minus, P is plus. As you can see from the table, there is now a nice simple coding, which highlights the extreme values:
Examination results for three subjects at three types of school

	French	English	Maths
Public School	·	·	·
Grammar School	—	—	·
Comprehensive	+	+	·

Now what can you say about these results?

TASK 13.3   Rerun the table in Task 13.2 replacing the table values with codes as described above.

It would be simplest to do a manual stem-and-leaf, and calculate the values for Q_u and Q_l from that. The computer can then work out the boundaries for the coding symbols:

```
10 INPUT QU,QL
20 IQR = QU - QL
30 OP = QU + 1.5*IQR
40 FP = QU + 3*IQR
```

(and similarly for the low values).

A print statement such as:

```
150 IF A(I,J) > QL AND A(I,J) < QU THEN PRINT "."
```

will print out the symbol for values falling within the box. You would require six more statements to cover the full range of symbols – tedious, but necessary.

13.2 POLISHING THE TABLE

Coding the table is a long step forward in reducing the confusion caused by the raw table of numbers. But you probably noticed that there was something else that should have been dealt with – an entry in the table may be coded as 'big' (+ in the synbol scale), but be in a big category. In the exam. results example 'comprehensive' was the biggest group, so maybe the + is only to be expected. What is needed is some way of spotting whether a value is 'big for its size' relative to the total value of the entries in its row and column. A method for doing this is known as 'median polishing'. It aims to strip out the row and column effects, and leave bare the real effects.

To illustrate the method, let's go back to the example or exam. results:

Examination results for three subjects at three types of school

	French	English	Maths
Public School	763	1205	996
Grammar School	98	205	543
Comprehensive	3431	2641	1901

The overall median for the entries in this table was found previously. 996.

First polish   Strip off the overall median from the table values.

	French	English	Maths
Public School	−233	209	0
Grammar School	−898	−791	−453
Comprehensive	2435	1645	905
			996

In the margin you add back the stripped-out value, 996 in this case.

Second polish   Strip out the remaining row medians.

First calculate row medians, and then strip out the row medians.
Calculate row medians:

	French	English	Maths	Median
Public School	−233	209	0	0
Grammar School	−898	−791	−453	−791
Comprehensive	2534	1645	905	1645
				996

Then strip out row medians:

	French	English	Maths	Median
Public School	−233	209	0	0
Grammar School	−107	0	338	−791
Comprehensive	790	0	−740	1645
				996

Third polish   Strip out remaining column medians.
First calculate column medians, then strip them out.
Calculate column medians:

	French	English	Maths	Median
Public School	−233	209	0	0
Grammar School	−107	0	338	−791
Comprehensive	790	0	−740	1645
Median	−107	0	0	996

Then strip out the column medians:

	French	English	Maths	Median
Public School	−126	209	0	0
Grammar School	0	0	338	−791
Comprehensive	897	0	−740	1645
Median	−107	0	0	996

You are finally left with this table of left-overs, or 'residuals' to give them their proper name. The method of stripping out medians makes certain assumptions about the way that the table is put together, mainly to do with additiveness – that the results are the sum of an effect for the subject and quite separately an effect of the type of school. This is a useful simplification to make, but you should be on the lookout for situations where this may not be the case.

You may have noticed that I have kept a record of row and column medians, and the overall median. Using these values it is possible to re-create any table entry:

$$\text{Entry} = \text{overall } M_d + \text{row } M_d + \text{column } M_d + \text{residual}$$

Note that the residual values would work out differently if you changed the order of the calculation. But the significant residual values should still show up.

The residuals can also be coded, using the symbols defined above: (You may like to draw up a stem and leaf plot of the values at the last step, just to check that I've got it right.)

	French	English	Maths	Median
Public School	−	·	·	0
Grammar School	·	·	+	−791
Comprehensive	++	·	=	1645
	−107	0	0	996

Now what do you think of the different forms of schooling, and their effect on exam. performance? Compare this with the original coded table. Remember that nothing you see here *proves* that one school is better than another. There is certainly a hint that Comprehensives are good at French. But you are engaged here in *exploratory* analysis. Proving a point, deciding if the balance of the probabilities is pointing unequivocally in one direction, is an aspect of confirmatory, not exploratory statistics. In this book we look, in the main, at exploratory statistics, leaving the confirmatory material for another time.

106

Stripping out the median need not be a purely mechanical process. For example, the median of row 2 (−898, −791, −453) is correctly given as −791. But with these three values perhaps 650 would be more central. Be adventurous!

TASK 13.4 Perform the median polish on your table in the last task. Use the computer where appropriate.
    Use *the program to investigate the effect of:*

(a) Changing the order of stripping
(b) Taking values slightly different to the median, if it seems more appropriate.

The program should read the data into an array, which will be gradually altered to show the effect of the median stripping:

```
10 DATA 763, 1205, 996
20 DATA  98,  205, 543
30 DATA 3431, 2641, 1901
35 DIM A(4,4)
40 FOR I = 1 TO 3
50    FOR J = 1 TO 3
60          READ A(I,J)
70    NEXT J
80 NEXTI
```

The print statement will be similar, replacing READ with PRINT. To strip out the overall median, I would suggest the following.

Use a pencil and paper stem and leaf plot to work out the median, and INPUT it to the program, to be stored in A(4, 4). Subtract this value from each entry.

```
100 INPUT "MEDIAN"; A(4,4)
110 FOR I = 1 TO 3
115    FOR J = 1 TO 3
120          A(I,J) = A(I,J) - A(4,4)
130    NEXT J
140 NEXT I
```

Then print out the table again.

Strip out the row medians in a similar fashion, but this time you can probably read off the values from the table on your screen:

```
200 FOR I = 1 TO 3
210    PRINT "MEDIAN FOR ROW"; I;: INPUT (I,4)
220    FOR J = 1 TO 3
230          A(I,J) = A(I,J) - A(I,4)
240    NEXT J
250 NEXT I
```

Print out the table, including the row medians, making sure it is giving you the right answer. Finally, strip out the column medians, which is similar to row stripping, but with the Is and Js switched. The coded plot of the residuals is probably a job best done by hand.

Remember to check carefully that your program is doing what you meant it to. It is very easy to fall into the trap of believing that the program is producing the results that you want. The computer only produces what you tell it. Wanting and telling are not always the same thing, so check, check, check!

CHAPTER 14

CORRELATIONS

14.1 RELATIONSHIPS

Is smoking related to lung cancer? Is inflation linked to the supply of money? Does the price of a house depend on its number of bedrooms? All of these questions look to a relationship between two sets of data:

$$
\begin{array}{ll}
\text{[smoking} & \text{lung cancer]} \\
\text{[money supply} & \text{inflation]} \\
\text{[number of bedrooms} & \text{house price]} \\
X & Y
\end{array}
$$

Underlying these statements is a belief in cause and effect, even sometimes a desire to 'pin the blame'. In the case of house prices, it is more a case of trying to explain how the price varies with the number of bedrooms. In all of these examples it was:

$$
\text{[CAUSE, EFFECT]} \\
[X, Y]
$$

The cause is labelled X, and the effect is labelled Y. It is of the greatest importance to know what you want from your data, what is the supposed cause, and what you take to be the effect. It would be just as valid to say: 'Inflation brings about (causes) increases in the money supply' as it is to say 'increases in the money supply cause inflation'. The same data could support either idea, so you must sort out your ideas about relationships before getting too involved in the statistics.

14.2 PLOTTING RELATIONSHIPS

'Inflation is caused by increases in the money supply' is the central belief of the economic doctrine known as monetarism. The effect is not immediate; increases in money supplied by the Central Bank is reflected, it is said, in the inflation figures that come later. To test this belief, here are some figures from the UK Government publication *Economic Trends*. The table below shows the inflation rate half a year after the corresponding money supply figure.

Year 1/4	M3% per 1/4	Year 1/4	Inflation per yr	Notes
1980 1	3.0	1980 3	16.5	M3 money supply
2	5.7	4	15.6	is the % increase
3	4.6	1981 1	13.1	per quarter (1/4).
4	4.2	2	11.8	
1981 1	2.1	3	11.3	
2	4.4	4	11.7	Inflation is the %
3	4.7	1982 1	10.7	increase in prices
4	1.7	2	8.9	over the previous
1982 1	2.6	3	8.0	year, 4 × 1/4.
2	2.4	4	6.5	
3	1.8	1983 1	5.5	Inflation is 'lagged'
4	2.2	2	4.4	by half a year.
1983 1	4.1	3	4.4	
2	2.6	4	4.4	
3	0.7	1984 1	4.6	

Source: *Economic Trends* (UK Government publication)

Plotting these values can be done on graph paper, but remember the 'hints for plotters' in Chapter 7. In Fig. 14.1 you can see what these values look like plotted out.

The graph shows a great deal of scatter – for this reason it is sometimes known as a scattergraph. But what is fairly clear is that there is a definite trend – higher values of money supply seem to go with higher values of inflation. The graph is the first vital step in understanding relationships. Note that the 'cause' (M3 money supply) is plotted along the bottom, or *x*-axis, with the 'effect' (inflation) plotted up the side, on the *y*-axis. This is a widely accepted convention in graph plotting.

TASK 14.1   Investigate the relationship between house price and number of bedrooms.
Select *a random sample of 20 houses from your data set, and plot them on a scattergraph, by hand. (More than 20 would take too long.)*

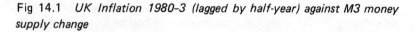

Fig 14.1   *UK Inflation 1980-3 (lagged by half-year) against M3 money supply change*

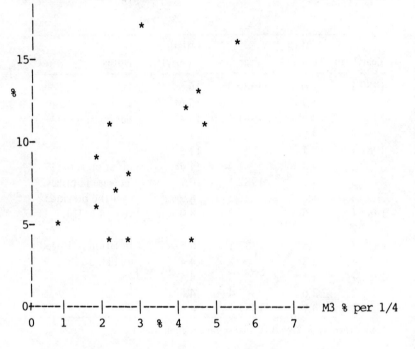

14.3 SCATTERPLOTS ON THE COMPUTER

Getting a scattergraph on the computer has several advantages.

(1) It is quick.
(2) You can usually see at a glance if some of the data values are incorrect, when one value is way out of line with the others.
(3) You can redraw the scatterplot with alternative scales.

A problem with computer plotting is that the programming is not very easy, especially if you want a 'nice' plot complete with scales in even steps, yet well spread out to fill the screen. Another problem is that the screen can normally plot only single characters. This means that you have a very coarse 'grid', which can only work at the expense of severe roundoff.

As graphics features vary so much on different computers, the rest of this section is by way of indication only. If you can spare the time, learn how your own micro produces high resolution graphics, and try to improve on my suggested program.

TASK 14.2   Plot the data for the last Task on the screen of your computer. Use the following suggested program, or (preferably) make use of the high resolution feature on your own micro.

Fiddle with the scales to produce 'stretched' and 'squeezed' plots.

Here are some programming hints for this task, using the money supply figures as an example.

Data from the money supply example can be entered onto DATA lines:

```
10 DATA 3.0,16.5,5.7,15.6,4.6,13.1,4.2,11.8,2.1,11.3,4.4,11.7
20 DATA 4.7,10.7,1.7,8.9,2.6,8.0,2.4,6.5,1.8,5.5
30 DATA 2.2,4.4,4.1,4.4,2.6,4.4,0.7,4.6,-1,-1
```

In the following program, I am assuming that the screen has at least 40 characters across and 32 lines down, so I clear a space 30 × 25 that will fit within that – this leaves space for axes and titles.

50 DIM A(31,26)   :31,26 is 30 × 25 plus 1 more all round

'Scale factor' is a number which multiplies up the value read. For example, if M3 ranges from 0.0 to 6.0, and the x-axis has 30 spaces, a scale factor of 5 will fill the space exactly. When you use the program, you will soon learn how to adjust the scale factor to get a graph which fits nicely:

```
60 INPUT "SCALE FACTOR FOR X"; SX
70 INPUT "SCALE FACTOR FOR Y"; SY
```

In the program segment below, the lines 100 to 160 read the data pairs, and record where they are on the grid. Because it is possible to get more than one value at a grid point, line 150 counts how many have been found at that point. Also if the scales have been badly chosen, then points may fall outside the grid. To accommodate this, you can imagine a 'border' all around; into this border all extreme values are placed: (A(0,*),A(31,*) bottom and top, A(*,0) and A(*,26). This explains the 'extra' subscripts in the DIM statement.

```
100 READ X,Y
110 IF X < 1 THEN 490
120 I = X * SX
130 J = Y * SY
141 IF I < 1 THEN I = 0
142 IF I > 30 THEN I = 31
143 IF J < 0 THEN J = 0
144 IF J > 25 THEN J = 26
150 A(I,J) = A(I,J) + 1
```

Having counted how many values there are at each grid point, the job of presenting the results on the screen still remains. What I have done here is to start at the maximum value of Y, which is 26, and print out all the X values at that grid level, spaced across. A$ is built up, first with an axis line (510). P$ is a character string representation of the number held at A(I,J). If there are no values (A(I,J) = 0) then we want to print a blank (540); if there are more than 9 values, that is two digits, print "*" (550). The values of P$ are added onto A$ like beads; when the line is full, it can be printed out. Line 490 gives a scale value for reference.

```
490 PRINT 25/SY
500 FOR J = 26 TO 0 STEP -1
510   A$ = " !"
520   FOR I = 0 TO 31
530     P$ = CHR$(A(I,J)+48)
540     IF A(I,J) = 0 THEN P$ =" "
550     IF A(I,J) > 9 THEN P$ = "*"
560     A$ = A$ + P$
570   NEXT I
580   PRINT A$
590 NEXT J
```

Finally we need to print a scale along the bottom:

```
650 PRINT" 0+-------------------------------";30/SX
```

In use the program will give you a scatter of points which will give you an indication of how good the relationship is. (You will probably find you need to do a bit of 'tuning' to get the scale right. If the values are big, e.g. house prices, then the scale factor is a number like 0.0005 or 5/10 000. You will know when you have a fit when there are few if any values in the border.)

One use you can make of this plot is to look in more detail at a particular area by enlarging it on the screen. Say you wanted to see what happened to M3 between the values of 2.5 and 4.0. To make values of M3 less than 2.5 drop into the border insert

```
115 X = X - 2.5
```

A suitable scale factor will cut off the values above 4.0.

14.4 FITTING A FORMULA

The relationship between two variables can be judged by the plot of the paired values. From this you can see whether there is *any* relationship, and also judge the quality of the relationship. The less the scatter, the better the relationship.

The formula

It is possible to work out a formula – 'fit a line' – to these data points on the scattergraph. The easiest and simplest line to fit is a straight one. But before fitting a straight line to your set of data pairs, look at Fig. 14.2 which reminds you of the equation of a straight line.

Fig 14.2

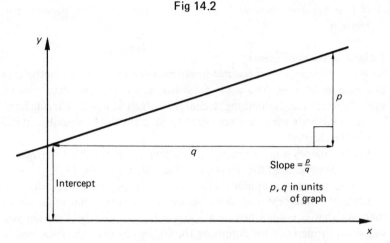

The *slope* of the line is measured as the ratio p/q. Any right-angled triangle drawn on the line will have the same slope, so long as p and q are measured in the same units as their own axes. If the line drops downwards to the right, the slope is taken to be negative; otherwise it is positive.

The *intercept* is the length cut off on the y-axis by the straight line. Note that this only works if the y-axis is drawn through the zero on the x scale.

The slope is given the label b, the intercept is given the label a and the formula for the line is written:

$$Y = a + b*X$$

a – intercept and b – slope are known as the 'coefficients' of the equation.

(You may be more familiar with this formula in the version:

$$y = m*x + c \qquad m - \text{slope}, c - \text{intercept}$$

which is the standard maths form. Statistics uses b for the slope and a for the intercept all the time, so please get used to this.)

Interpreting the coefficients

In relationship terms these two coefficients a and b, intercept and slope, can be interpreted as:

a (intercept) is the value of *Y*, when *X* = 0. When the effect of the 'cause' is zero, *a* gives the 'at rest' value of *Y*.

b (slope) is the 'gearing' of the relationship, and is the single most important value. If *X* is increased by 1 unit, *Y* changes by *b* units.

if *b* = 1, then there is a 1:1 relationship between *X* and *Y*.
if *b* > 1, *Y* increases faster than *X*.
if *b* is negative, then as *X* rises, *Y* falls (this is called 'negative correlation').

Estimating the coefficients
Given a scattergraph, it is possible to estimate the values of the coefficients in a number of ways. You could guess the values from the scatterplot by eye. There are also 'smoothing' techniques which you will be trying later. Formula methods which are very easy to program will be described at the end of the chapter.

On the money supply example, it is just about possible to imagine a straight line which best fits the points. This is shown in Fig. 14.3.

If you use the computer scattergraph, you may have difficulty pinpointing the intercept and slope values. Don't worry too much about that. All that is needed here is a rough estimate. In a later section you will discover methods for improving the fit. In any case, the data used in this and in most investigations is of uncertain quality. A value for the slope correct to the fifth or sixth decimal place has got to be fraudulent!

TASK 14.3   Use the scattergraph to estimate the slope and intercept of the relationship between house prices and number of bedrooms.

To give you some idea of what sort of answers to expect: if the median house price was £50 000, I would expect a value of about £10 000 for *b*, and about £20 000 for *a*.

You can ask yourself whether these make sense by making a statement such as 'each extra bedroom adds £10 000 on average to the price', *or* 'a house with no bedrooms (vacant plot) would cost about £20 000'.

Check your values for reasonableness. Unless you believe your coefficients have sensible values, further analysis is pointless.

14.5 TESTING THE FIT: RESIDUALS

Estimating a formula for the straight line is a way of summarising the data pairs. Instead of all the *X* and *Y* values, just *two* coefficients, *a* and *b* can be given. The formula:

$$Y = a + b*X$$

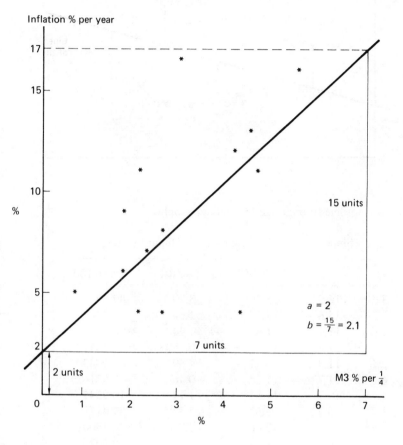

Fig 14.3   *UK inflation 1980–83 (lagged by half-year) against M3 money supply change*

allows you to *predict* a value of Y, for any given value of X. Of course, the scatter of the actual values means that this prediction is not always precise. The differences between the predicted and actual values of Y are called the 'residuals'. Note that the residuals are only measured on Y; it is assumed that X values have no left-over components. Figure 14.4 shows how residuals are measured.

As well as showing the values on a scattergraph, it is an easy job to calculate the residuals with the aid of your computer.

Fig 14.4

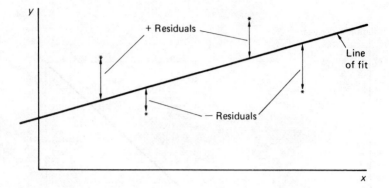

Results for line fit on inflation v. money supply:

	Slope = 2.1	Intercept = 2	[Input by user]	
X	Y	YPRED	Residual	Ratio*100
3	16.5	8.3	8.2	49.697
5.7	15.6	13.97	1.63	10.4487
4.6	13.1	11.66	1.44	10.9924
4.2	11.8	10.82	0.980	8.3051
2.1	11.3	6.41	4.89	43.2743
4.4	11.7	11.24	0.46	3.93162
4.7	10.7	11.87	−1.17	−10.9345
1.7	8.9	5.57	3.33	37.4157
2.6	8	7.46	0.540	6.75001
2.4	6.5	7.04	−0.54	−8.30769
1.8	5.5	5.78	−0.28	−5.0909
2.2	4.4	6.62	−2.22	−50.4545
4.1	4.4	10.61	−6.21	−141.136
2.6	4.4	7.46	−3.06	−69.5454
0.7	4.6	3.47	1.13	24.5652

Residual is Y (actual) − Y (predicted)
Ratio is Residual/Y (actual)

It is usual to express Ratio as a percentage.

Both residual and ratio can be used to examine the effectiveness of the formula in predicting the actual value of Y. The ratio would be useful if the proportional change in Y was the main feature of interest.

Residuals can be used in *two* main ways.

(1) Spot the maverick. If one value has a much larger residual than the rest, then it deserves closer attention. Can you spot the maverick in money supply? When did that happen? (Check back on the original data.)

(2) Suggest an alternative fit. Looking at the residuals in the table above, it is clear that the formula gives a prediction which is too small for a big Y, and too big for a small Y. To improve the fit, I should take a steeper (bigger) value for b. A more systematic method for improving a and b will be explained later.

TASK 14.4 Using the rough estimates of slope and intercept calculated in the last Task, work out the predicted values of Y, the residuals and the ratio between residuals and Y (actual).

Print out the residuals, and comment on the extremes.

Plot the residuals using your computerised scattergraph, and comment on any pattern.

Vary the values of a and b, and investigate the effect on the residuals of these changes.

Calculating the residuals is very simple on the computer:

```
10 INPUT A,B
20 READ X, Y
30 YP = A + B*Y          YP - predicted value of Y
40 RES = Y - YP          RES - residual
50 RATIO = RES/Y *100
60 PRINT X, Y, YP, RES, RAT
70 GO TO 20
```

Plotting can be carried out with a little modification to the original scattergraph program:

```
110 I = X*SX
120 J = (A + B*X - Y)*SX + 12
```

The +12 is to allow for negative residuals. You will have to remember that the zero of the scale is in the middle of your screen.

14.6 IMPROVING THE FIT: BETTER VALUES FOR a AND b

In the last task a method for improving the values of a and b was hinted at. In this section the method will be put on a more formal basis.

Since residuals tell us how good the fit has been, reducing the total value of residuals by varying a and b will give an improved fit. But since some residuals are positive and some negative, you have to add up the *absolute* (positive only) value of the residuals. If not, then there is a strong possibility that the positives and negatives would cancel each other out. The process can be described as follows:

(1) Estimate values for a and b (from the scattergraph).
(2) Calculate values of residuals using these values.
(3) Add up the absolute values of the residuals, and note their sum.
(4) Vary a until the smallest sum of absolute residuals is found.

(At this stage you may decide that a few of the values are giving residuals which are much too big, far outside the normal run. In that case the best thing to do would be to trim them off, especially if you have reason to believe that they are unusual.)

(5) Using the value of a found in (4) change the value of b, until the smallest absolute sum of residuals is found.

You may repeat steps (4) and (5) to get even greater accuracy, but don't be unrealistic. If the data is correct to the nearest £1000, then values for the coefficients correct to the nearest £10 will do fine. Note that this is not the only way of teasing out values for the coefficients a and b, the intercept and the slope. It is certainly not the most efficient method, but it serves a purpose in helping you to understand how a line is fitted to a scatter of points.

TASK 14.5   With the help of a computer program, find reasonably accurate values for a *and* b *using the method described above.*

14.7 IMPROVING THE FIT: METHOD OF LEAST SQUARES

The technique used in the last section attempted to fit a line which minimised the absolute deviations from that line. In this section an alternative method is described, which first squares the residuals, and then minimises the sum of these squared residuals – hence the name 'method of least squares'.

Squaring gets rid of the minus signs, which is a help. It also gives a different weighting to the values, emphasising the distant values at the expense of the near-in values. Try the least-squares method to see if it gives greatly different coefficient values.

TASK 14.6   Replace absolute residuals with the squared residuals in the program used in the last task. Repeat the exercise for finding the best a *and* b *using the minimum sum of squared residuals.*

Compare *the results with those found using the previous technique, and comment on the ease of finding the best* a *and* b *using this method.*

14.8 METHOD OF LEAST SQUARES BY FORMULA

The method of least squares described in the last section was a hit-and-miss technique – you guessed at values for slope and intercept and then improved them. But there also exists a formula method for the calculation which gets you to the answer directly. This method is widely used by statistical packages on computers. Some pocket calculators with statistical functions will work a and b out for you at the touch of a few buttons. It is also quite easy to program this method.

What follows is just a bald statement of the formula, not a proof, or a derivation.

For the slope the formula is:

$$a = \frac{\Sigma Y - b\Sigma X}{N} \qquad N \text{ – Number of data points}$$

$$b = \frac{N\Sigma XY - \Sigma X\Sigma Y}{N\Sigma X - (\Sigma X)^2}$$

TASK 14.7   Write a program to calculate the value for the slope and intercept of a straight line fitted to a set of points, using the method of least squares formula.

Programming hints: This program is not too different from the calculation of arithmetic mean. You calculate not just the sum of X, but the sum of Y, sum of X squared, and the sum of X times Y.

In the following program, these variables will be used:

SX – sum of X
SY – sum of Y
XX – sum of X squared
XY – sum of X times Y

```
 5 N=0:SX=0:SY:0:XX=0:XY=0
10 READ X,Y
20 IF X < -999 THEN 100
30 N = N + 1
40 SX = SX + X
50 SY = SY + Y
60 XX = XX + X*X
70 XY = XY + X*Y
80 GOTO 10
100 REM calc b and a
```

```
110 B = (N*XY - SX*SY)/(N*XX - SX*SX)
120 A = (SY - B*SX)
130 PRINT A,B
```

14.9 CORRELATION COEFFICIENT

Another widely used formula calculation is described in this section. You will have seen from your scatterplot that the points did not all lie on a perfect straight line. This situation is highly variable: sometimes the points are close to the line, sometimes you will find a huge scatter. In earlier tasks you judged how good the relationship was by looking. In this section you will work out a number which can be used to gauge how effective the straight line fit has been in explaining the scatter.

The measure which indicates this goodness-of-fit is called the 'correlation coefficient'. The formula for it is:

$$R = \frac{N\Sigma XY - \Sigma X \Sigma Y}{\sqrt{[N\Sigma X^2 - (\Sigma X)^2][N\Sigma Y^2 - (\Sigma Y)^2]}}$$

Again, no explanation is given for this formula – that can be found in many textbooks. The result of this calculation is a number between 0 and 1. One indicates a perfect fit, zero indicates total scatter. You always end up somewhere in between, so this value provides a basis for comparing the fit of different sets of data. The correlation coefficient also has a sign + or − which follows the sign of b, the slope.

TASK 14.8 Write a program to calculate the correlation coefficient, using the formula given above.
Alter your data set to try out the effect on R.

The program is just a small extension of the previous task

YY – sum of Y squared

```
75 YY = YY + Y*Y
```

```
140 R = ( N*XY - SX*SY) /(SQR( (N*XX - SX*SX)*(N*YY - SY*SY))
150 PRINT R
```

This has been a long chapter, with a large number of tasks, some involving a lot of programming. But the topic of fitting a relationship is important enough to merit all this attention. Fitted relationships are used in a wide variety of applications, not always with a real understanding of what is going on. There are many more techniques and formulas used in curve fitting, which you may come across. There is transformation of the data, which enables you to try out non-linear relationships. There is multiple regression analysis, where you use not one, but a number of variables to explain the effect under investigation. In this chapter I have introduced you to some of the ideas of fitted relationships.

CHAPTER 15

TIME SERIES

There are many series of data which are followed with great interest by business analysts, stockbrokers, politicians and even sports fans. You must have heared of some of them: On the Stock Market there is daily reporting of the share-price index. For the cost of living there is the once-a-month reporting of the latest Retail Price Index. The weather and football results provide a stream of daily, weekly and seasonal figures to chew over. But all too often there is a great deal of over-reaction to the latest figures. People fail to put the numbers in a historical context and ask, 'Is this value really different, or is it just a freak?' This chapter will show you ways of exploring time series data with the aid of your microcomputer. This chapter differs from all the previous chapters in one major respect – I will not be asking you to make use of your house price data set. The reason for this is practical – to get a reasonable time series of data on house prices would require a couple of years, and I am sure you don't want to wait that long before starting on this chapter! So the data here will be examples taken from published data sources.

15.1 WHAT IS A TIME SERIES?

In many ways time series analysis is like the relationship investigations of the last chapter. We are looking for a relationship between some measured value (say a price index) and *time*. The methods of the last chapter could be applied, but time series have *two* characteristics that set them apart:

(1) Data are always reported at fixed regular intervals.
(2) One value follows on from the last, is connected together.

For these reasons, time series need to be looked at using techniques specific to them.

The reasons for wanting to analyse time series are: (a) to see what has happened and to try and understand why; and (b) to try predicting (forecasting) what is going to happen.

Accurate prediction has been a desire of mankind from the dawn of time. There is a touching belief, not limited to computer specialists, that accurate computed prediction will soon be a possibility. Fat chance! The track record of forecasters, especially for long-term forecasts, is poor. What we should look for in time series analysis is understanding of what is going on. In this way we can hope to make more educated guesses about the future.

15.2 DATA

The data for this chapter follows on from the example on money supply used in the last chapter. For a government which is committed to the monetarist philosophy, the changes in M3 are of the utmost importance. (M3 is money defined in a very broad way to include non-cash items like credits.)

Example 1 M3 money supply. UK percentage increase per month:

Data for Jan 1982 to June 1984 (30 months).

	J	F	M	A	M	J	J	A	S	O	N	D
1982:	1.4	0.7	0.5	1.2	0.5	0.9	0.1	1.0	0.9	1.4	0.2	0.1
1983:	1.4	0.7	1.1	2.1	0.7	1.0	0.5	0.5	0.3	1.0	0.1	1.1
1984:	0.7	0.1	1.4	0.4	0.9	2.0						

Example 2 M2 money supply' UK percentage increase per month:

Data for Jan. 1982 to June 1984 (30 months).

	J	F	M	A	M	J	J	A	S	O	N	D
1982:	0.1	−0.6	1.0	0.8	0.2	1.2	1.3	−0.2	0.5	1.9	0.2	1.2
1983:	−0.1	0.4	1.5	1.4	0.5	1.0	1.0	−0.5	0.1	0.7	0.2	2.8
1984:	1.0	***	1.3	1.9	0.5	1.5						

*** indicates a missing value

Source: *Economic Trends* (UK Government Publication) July 1984.

M2 is similar to M3, except it is more narrowly based.

15.3 **PLOTTING THE DATA**

Plotting time series can be done very easily on the computer. Because there is one value only in each time period, and because these time periods are equally spaced, you can plot with one value per print statement. But because it is much easier to print down the screen, rather than across, I am proposing a turnaround in the plotting method. Instead of following the convention and plotting across the page, I suggest you plot *down* the screen. The result looks like this (Fig. 15.1). (If you turn it around, it looks more conventional.)

Fig 15.1 *plot of raw data. M3 money supply percentage increase in months from January 1982*

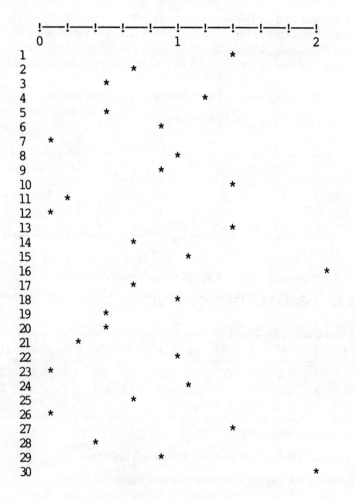

TASK 15.1   Using the data for M2 (Example 2 above), get your computer to produce a screen plot.

Programming hints: To allow the plot to be spread out to suit your screen, I have included line 50, which enquires about the scale factor.

```
10 DATA 1.4,0.7,7.0 .....etc
20 DATA ......................,-999
50 INPUT"SCALE FACTOR";SF
60 PRINT"    !----!----!----!----!----!----!----!----!"
70 PRINT"    0";TAB(5+SF)"1";TAB(5+SF*2)"2"
100 N=0
110 READ Y
120 IF Y = -999 THEN 200
130 N=N+1
140 PRINT N; TAB(Y*SF+5)
150 GOTO 110
200 STOP
```

A note about the data for M2: I have left you with two little problems here:

(1) Some of the data are negative, so you will have to adjust the scales to suit.

(2) *** indicated a missing value. What can you do about that?

Missing values are a real problem with time series. Whatever the reason for this one - strike, breakdown of computers or something else - you have to cope with the values as you find them. The usual thing to do is to 'interpolate' a value - estimate what it should have been, and use that.

To interpolate, we can use nearby values. Taking the median of the two values either side gives:

$$1.0 , 1.3 \quad \text{median } 1.15 \text{ say } 1.2$$

You may feel it would be better to take the median of four or six values.

15.4 SMOOTHING THE DATA

Just looking at the plot in the last section tells you very little, apart from the obvious - M3 values are highly erratic. So how can you see what is happening to the M3 money supply? The answer is to smooth out the bumps. Replace each value in the series with the median of itself and its neighbours:

Taking the first three values in the M3 series

$$1.4 \quad 0.7 \quad 0.5 \quad \text{median } 0.7 \text{ (2nd smoothed value)}$$

and the next group of three

> 0.7 0.5 1.2   median 0.7 (3rd smoothed value)

The first smoothed value can be found by inventing a previous value:

> 0 1.4 0.7   median 0.7

This process of smoothing is based on a median of 3, so it is sometimes called smoothing with a 'span' of three. The results look like this (Fig. 15.2). (Again you need to turn it around to see it in conventional terms.)

Fig 15.2 *plot of smoothed data from Fig 15.1*

```
       !—!—!—!—!—!—!—!—!—!
       0               1               2
  2                 *
  3                 *
  4         *
  5                   *
  6         *
  7                 *
  8                 *
  9                    *
 10                 *
 11   *
 12   *
 13           *
 14                      *
 15                      *
 16                      *
 17                   *
 18             *
 19       *
 20       *
 21       *
 22     *
 23                  *
 24             *
 25             *
 26             *
 27       *
 28                *
 29                *
```

Before getting you to write a program to smooth out your data, it is a good idea to look at the problem of finding the median value of three numbers. You have written a program before, in Chapter 5, which found the median of 100 values. The job of finding the median of three values seems so simple, when doing it in your head! But it turns out to be quite a ticklish little problem to program, so try your hand.

TASK 15.2   Given three numbers, write a segment of program which will return the median value.

(If you really get stuck on this one, here is a crude program to find a median of three):

```
500 REM INPUT VALS ARE X1,X2,X3; MEDIAN IS YM
510 IF X1=X2 OR X1=X3 THEN YM=X1
520 IF X2=X3 THEN YM=X2
530 IF (X1>X2 AND X1<X3) OR (X1<X2 AND X1>X3) THEN YM = X1
540 IF (X2>X1 AND X2<X3) OR (X2<X1 AND X2>X3) THEN YM = X2
550 IF (X3>X1 AND X3<X2) OR (X3<X1 AND X3>X2) THEN YM = X3
560 RETURN
```

TASK 15.3   Change the program in Task 15.1 to calculate and print out a smoothed span-of-three series for the M2 data.

Finding the three values to be smoothed is the first change. Once that is done, the program segment in the last Task will give the median to be printed out. Let the three values to be smoothed be X1, X2, X3. When a new value of Y is read in, these values have to be updated:

115 X1=X2: X2=X3 : X3=Y

and then you can pick the median:

118 GOSUB 500   which finds YM for printing out.

But a problem exists at the start, so you have to set up X1, X2, X3:

105 X1=0: X2=0 : READ X3

(You should also do something about the last value, which as you have probably spotted is missing in my diagram.)

15.5 DOUBLE SMOOTHING

Smoothing using a span of three has reduced the amount of scatter in the points, but looking at the smoothed diagram it is still not too clear what has been happening to M3. So if one smoothing operation failed to do the trick, then do another one.

128

Double three-span smoothing means you smooth once on the original data, and then a second time on the smoothed data. Figure 15.3 is a plot of M3 after a second lot of smoothing.

Fig 15.3 *plot of double smoothed data from Fig. 15.1*

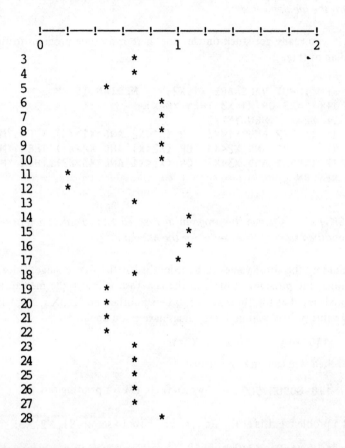

TASK 15.4 Double smooth with a span of three the data for M2 money supply. Plot out the results.

Comment *on any pattern you detect, and how it compares with single smoothing.*

There are two smoothing stages in this program. The first is like the previous single smooth, but instead of printing it out, you store it in an array:

    140 S(N) = YM

Don't forget to DIMension S(30).

The second stage is like the first, but based on values in S(I).

```
200 X1=0: X2=0: X3=S(1)
210 FOR I = 2 TO 30
220   X1=X2: X2=X3: X3=S(I)
230   GOSUB 500
240   PRINT I-1;TAB(5+SF*YM)
```

15.6 MULTIPLE SMOOTHING

If you can smooth twice, then there is no reason why you shouldn't do so again. Many different smooths can be tried.

Different spans

Five, seven even nine spans can be tried. The odd numbers mean that the median is found opposite one of the original data points. Even number spans may be used, but they present a problem:

$$X \quad X \quad X \quad X$$
$$1 \quad 2 \quad 3 \quad 4$$

With four values, the median will fall between 2 and 3. In order to make the smoothed value line up with the original value, a further smoothing of span is carried out. Even smoothing is normally done like this in two stages, a four span and a two span.

Combinations

A span of three could be followed by a span of five. If that doesn't work then try another span of five.

Other forms of smoothing

There are many other techniques of a mathematical and graphical nature which have been invented to smooth out rough data. One area where a lot of work has been done is in Actuarial Science (Insurance Analysis). Actuaries have developed a whole range of processes for smoothing data, or 'graduation' as they call it.

15.7 RESIDUALS: THE 'ROUGH'

Having smoothed out the data, there is a great temptation to leave it at that. But in all the previous work on relationships we have always looked at the residuals, the left-overs when the relationship has been fitted. So here we go again.

TASK 15.5 *Calculate and plot out the 'rough' from single and double smoothing of the values for M2.*
   Comment *on any pattern found.*

You can even look for pattern in the 'rough', again making use of the smoothing process. This may seem an outlandish thing to do, but it is not as silly as it sounds. Suppose there are two main trends in the data. The smoothing may reveal one of them, especially if it is dominant. But only by looking at the rough, the residuals, will you be able to pick up the second. So the rough is always worth a second look.

Remember! the purpose of a diagram is to let you see what you never expected to see. *Keep exploring your data*!

TASK 15.6 *Smooth the roughs from Task 15.5, plot them out, and comment on any pattern you discover.*

PART VI
FURTHER THOUGHTS

This part adds a few further ideas to what has already been learned. No more tasks, just some reading.

CHAPTER 16

MISTAKES, ERRORS AND SURVEYS

Congratulations! If you have worked your way through all the tasks up to this point, then you have completed a major undertaking. Relax! From now on there are no more tasks for you to do, only some reading, but important reading, nevertheless. The task, if you like, is to read the following pages, and note their contents.

Admitting that you can make mistakes is not a sign of weakness, rather it shows a clear-sighted view of reality. Everyone makes mistakes from time to time. Acknowledging this fact of life means that we can plan to prevent mistakes, or at least be aware of mistakes when they happen.

A *survey* is a data collection exercise, usually with a specific answer in view. Examples would be your house price survey or a public opinion poll prior to an election. The ten-yearly Census of Population is an example of a survey with general purposes in mind. There have been many embarrassing failures of surveys to produce correct results. The reasons for this failure fall under two headings – mistakes and errors.

A *mistake* is some act or omission which could have been avoided. Saying that $2 + 3 = 6$ is an obvious example of a mistake. Saying that the median price of a dwelling was £51 000 could be a mistake if your survey only covered houses, and left out dwellings such as flats.

An *error* is an inaccuracy which can arise even though no mistake has been made. All data-recording and calculation involves error. As you saw in Task 11.4, the median of repeated samples of 10 houses was not always the same, even though they were all drawn from the same set of data. This variability means that any calculation is only approximate, and is subject to error. Another source of error is due to round-off on the computer or calculator, because of the limited number of significant digits held.

To summarise:

(1) Mistakes are to be prevented if possible; if not, they are to be detected and corrected.

'Failure to avoid mistakes and control errors has caused many surveys to go astray.'

(2) Errors are to be estimated and controlled to within acceptable bounds.

16.1 HOW TO PREVENT OR CORRECT MISTAKES IN YOUR SURVEYS

'Be careful – don't make mistakes' is the unhelpful advice given by generations of teachers. Yet mistakes continue to be made. One reason for this is that teachers seem to value speed above accuracy. In real situations it is far better to get the right answer slowly, than to get a whole lot of answers, some of which are wrong. Even more important is to be confident that you have the right answer – you can be confident because you have checked it, or be confident because you have calculated it in a different way, for example. This may seem like stating the obvious, but it never ceases to amaze me how those in authority simply blame their clerks for slackness or carelessness whenever mistakes happen. They cannot imagine that their own system can create mistakes.

Practical steps that can be taken to avoid mistakes include:

1. Clear definitions, good communications
Make quite sure you know what is to be measured by your survey, when and by whom. Make sure that everyone who needs to know is kept in the picture.

2. Neatness and clarity
Simple things like the way numbers are written or printed can be crucial. There are a number of problem areas:

O,0: Is it letter Oh or number zero? On the typewriter and computer keyboard they are near to each other. More confusion is caused by mixing these two characters than anything else.
1,I 5,S: Number one, letter I; number five, letter S cause mix-up problems.

The way numbers are printed can cause problems too. With a line printer or dot matrix printer there can be difficulties in reading the numbers correctly. Long numbers can easily be misread. Is 58134671 approximately five million, 58 million or 581 million?

Many surveys, especially those based on questionnaires, make use of preprinted forms. A well designed from can cut down on mistakes very significantly. If a number is always to be written in a particular place on a page, it avoids confusion and prevents mistakes. Using lined analysis paper also helps.

Orderly, well-spaced calculations are not just a good way of helping you to get the right answer. They also build confidence; one look is enough to confirm that this is a person who knows what he or she is doing.

3. Checking

There is no limit to the amount of checking you can do. The more you check, the more your confidence grows in the results.

Check for reasonableness: always look at the numbers to see if they fall within sensible bounds. A house with 50 bedrooms should ring alarm bells. These checks can be built into your computer programs.

Repeating the calculations is a dull method of checking. Look for alternative ways of getting the same answer, and check that way.

Check-sums are widely used in commerce to detect mistakes. For example, if you know that there are 100 houses in your data set, you can check your stem-and-leaf plot by adding up the frequencies in each interval, to see if they still add up to 100.

Check-digits are a specialised form of mistake-trap. In this age of computerised account numbers, book numbers and social security numbers, getting these numbers right is of the greatest importance. Each number is extended, by adding a 'check-digit'. This can mean that there is only one chance in ten, or if two digits are used, only one chance in a hundred, that your bank account number will be misquoted.

4. Controlling the surveyors

If you are using other people to help you collect the data, there is always a problem – how to prevent careless or skipped work. This is true whether your assistants are unpaid volunteers, or paid employees. How can you deal with this problem? Do you:

(1) Believe everyone is honest, until you catch someone, then dismiss them, or
(2) Believe that everyone will cheat if they have the chance, so let it be known that you operate continuous checks?

A useful method of spotting cheating is to look for data which is *too* good. It is very difficult to invent data which has the right amount of variability. If the inter-quartile-range drops suddenly, suspect cheating.

5. Avoid manual techniques, use electronic aids

Noting numbers, transferring numbers, doing calculations, are all areas where mistakes can happen. The more these operations can be avoided the less chance there will be for mistakes to occur.

A good example of a design which reduces mistakes is the mark-sense form. The response of, for example, a student on a multiple choice test is recorded by the student marking a blob in a special grid as shown in Fig. 16.1.

Fig 16.1 *standard mark-sense response grid*

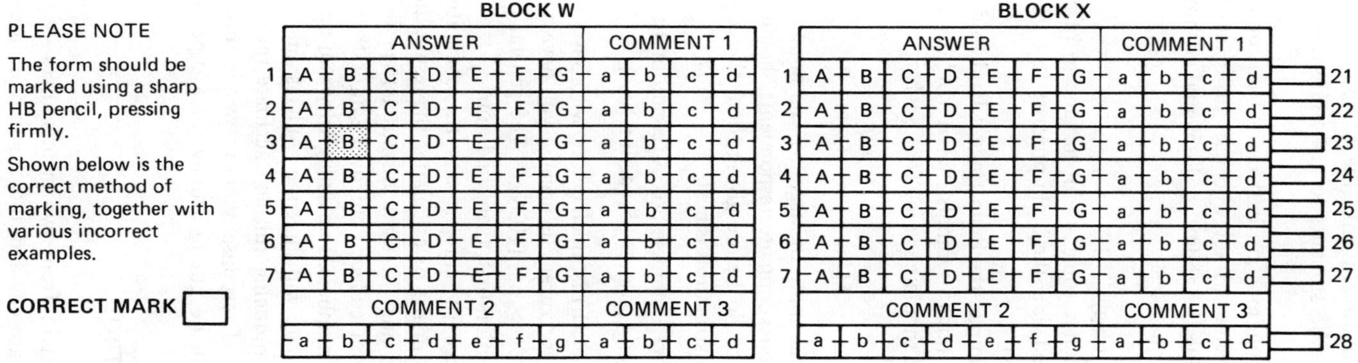

PLEASE NOTE

The form should be
marked using a sharp
HB pencil, pressing
firmly.

Shown below is the
correct method of
marking, together with
various incorrect
examples.

CORRECT MARK

This is then read directly into the computer, with no human inter-vention. It is up to the students to make sure that they fill the form in correctly.

Even simple calculations are best carried out on an electronic device. The cashier at the supermarket who uses his or her till to work out the change may be losing the ability to do mental arithmetic – but he or she will make far fewer mistakes!

6. Use a bit of psychology

Behavioural science has some useful insights to offer into our ways of perceiving things, and how the brain processes the information it receives (or thinks it receives).

To take one example, the brain finds it hard to distinguish between a large number of different symbols. Seven symbols, it has been found, is about the most that should be used. That is why the polished median table had seven symbols:

$$M = - \quad . \quad + \quad ++ \quad P$$

A similar rule should apply to diagrams – no more than seven categories.

Instructions to questionnaire respondents should be clear. It is better to tell them what to do before giving them the conditions for doing it.

Which of these instructions is best?

(a) If aged over 65, complete box A (condition first)
(b) Complete box A, if over 65. (action first)

Tests have found that (b) is much better, with fewer mistakes. The reason you may think (a) is 'correct' is because you have been trained, through BASIC, to say, IF X > 65 THEN . . . The general public finds this un-natural, preferring action first, then condition.

If you can engage the interest and the commitment of the people filling in the questionnaire or calculating the result, then you have a much better chance of getting the right answer. I tried to achieve this in this book, by making you collect your own data, rather than giving you my old cast-off sets of numbers. Explaining why you are collecting the data, and promising that you will share the results is time consuming, but it can pay dividends in producing full and accurate returns. If you are really keen to ensure a good response, you can even try bribery: 'Each correctly completed form will be put in a draw for a super weekend for two in romantic Paris'.

16.2 ERROR CONTROL

Errors in surveys stem from two main sources – variability you face in the data being measured, and the amount of information you gather.

If you were to restrict yourself to a single house type in a single suburb, the results would not vary greatly. If your survey covered a whole city, and all types of dwelling, you would expect a much greater range of values. The second survey is more variable than the first. This variability could be measured – the inter-quartile range of the first would be smaller than the city-wide survey. For a given size of sample, the results for the more variable data will be less reliable.

Whether you are trying to estimate the percentage vote for each candidate at an election, or the median price of a house in your suburb, the result of a sample survey will always be subject to an element of uncertainty. The amount of uncertainty will depend directly on the variability of the data to be measured.

Public opinion pollsters have developed specialised methods of controlling the effects of variability, but, in general, the only thing you can do is to take a bigger sample of data. The more data you collect, the more information you have. More information means less variability, less uncertainty. Unfortunately, the improvement in error control from each extra piece of data gets less. 'Diminishing returns' is the name economists give to this phenomenon – each extra unit producing marginally less than the previous one.

If you take a sample of data in order to draw conclusions about a wider population, then it is important that your sample should be drawn in a 'random' way. A random sample is defined as one in which each member of the population (all the items of interest) have an equal chance of being selected. Very often you must take your data where you can find it, but you must always question whether it is truly random and representative.

To summarise, to control error you should:

- know what variability exists in your population
- take a large enough sample to control that variability
- use random sampling methods.

CHAPTER 17

WHERE FROM?

WHERE TO?

17.1 HOW FAR HAVE YOU COME?

Statistics

This book is about statistics, so you will have learned something about that subject. You know how to collect data, because you gathered information in the real world on house prices. You understand that data can be classified in different ways, numerically or by categories. You have also gained practical skills in recording data, using a stem and leaf plot or a tally mark chart. These methods of recording data which you learned are also used for presenting data. In addition you learned how to present your data in graphical forms, using box and whisker plots and histograms. Learning how to carry out these methods of presenting data enables you to understand and interpret other peoples' statistical presentations.

But statistics does not stop at the collection and presentation of data; it also includes data analysis. In this area, you are able to carry out the calculations involved in finding middle values, spread values and skewness indicators in a set of data (you may not remember all the details of the calculations, but you know where to look if you need to know). But just being able to work out these measures is no 'big deal'. What you have also acquired is an understanding of what is being measured, how one form of measure compares with another, and above all when and how to apply these measures.

By learning methods of comparing data, and how to calculate index numbers, you acquired skills in exploring changes in sets of data which are measured at different times or places.

Looking for relationships in data took you into a more complex area of statistics. You know how to draw up a table, and 'polish' it so that you can begin to interpret what it is saying. Mathematical relationships can be explored using the plotting and formula methods you learned. The special case of time series required you to practise some special techniques of smoothing. As well as learning how to carry out the calculations to find a relationship, you gained an understanding of the implications and applications of the techniques.

To summarise: you have learned *skills*, especially in statistical calculations, but also in recording and presenting data. Much more importantly, you now understand what these calculations are for, how and when to apply them, and what they mean when you do. Quite an impressive result!

Computer programming
You may have arrived at this book with a scanty knowledge of programming. This is a common experience, even if you had previously been on an advanced course in BASIC programming. The difficulty as I see it is that learning without a specific purpose in mind is not very inspiring. In this book you have discovered a need for all sorts of program instructions – saving data on to a file, saving data in an external storage device, use of subscripted variables, use of screen graphics. If you were a little weak on programming before, doing the tasks has given you a breadth and depth of skill in this area.

Housing market
You must by now be something of an expert on the housing market in your survey area! There are many professionals involved in housing – solicitors, bank managers, building society valuers, surveyors, estate agents. The amounts of money involved in housing are massive – housing represents between a quarter and a half of all wealth in the economy. From your own point of view, buying or renting a house is almost certainly your biggest lifetime expenditure. For all of these reasons, knowledge and understanding of the housing market is a very useful asset.

Self-confidence
What I hope you have gained from this course above all else is *confidence*. Yes, you can explore data, on your own, for your own purposes. There are no right answers, only answers that you can believe in and stand by. You know you have the power to explore data – that is what your micro-computer is for. You also have a number of skills which you know you can use for looking at the data. You can do it!

17.2 THE NEXT STEP: MORE STATISTICS

Learning is a process which lasts a lifetime. There is no limit to how much you can know or need to know. The main source of information is books, so here is a selection of titles which you can refer to, and learn from.

Exploratory data analysis

This book is based for the most part on Exploratory Data Analysis (EDA). EDA is not a new subject, rather a fresh look at an old one – statistics. The label EDA was given to these techniques by John Tukey of Princeton University, who is also credited with most of the techniques such as box-plots and stem and leaf. Books about EDA are still scarce.

Understanding Data: An introduction to exploratory and confirmatory data analysis for students in the social sciences by Bonnie H. Erickson and T. A. Nosanchuk (McGraw-Hill Ryerson, 1977 and Open University, Milton Keynes, England, 1979).

This is the most readable account of EDA, although all the examples are slanted towards sociology. It is wordy, even long-winded, but if you want a much fuller explanation of most of the topics dealt with in this book, this is the one to look at.

Applications, Basics and Computing of Exploratory Data Analysis by Paul F. Velleman and David C. Hoaglin (Duxbury Press, Boston, Massachusetts 1981).

As its title suggests, this book is concerned with EDA and the computer. It gives full listings in both BASIC and Fortran for all of the techniques. This makes it a useful reference. In addition, there are very good explanations of the techniques of EDA, in some cases covering ground not mentioned in Erickson and Nosanchuk.

EDA Exploratory Data Analysis by John W. Tukey (Addison-Wesley, Reading, Massachusetts 1977).

Data Analysis and Regression Frederick Mosteller and John W. Tukey (Addison-Wesley, Reading, Massachusetts, 1977).

Graphical Methods for Data Analysis by Chambers, Cleveland, Kleiner and (Paul) Tukey (Wadsworth, California, 1983).

Taken together, these three books are the foundation of EDA. The first one went through several revisions in the 1970s. If you wish to see where the ideas of EDA came from you may wish to look at these books, but I should warn you, the style and content is most unusual. It is obvious that Tukey is an experienced practitioner and a most imaginative inventor of new techniques, but you may find his books difficult to work from. The book on graphical analysis is full of imaginative, programmable methods for exploring data.

Statistics in Society Open University Course MDST 242 (Open University, Milton Keynes, 2nd ed. 1984).

This course is available as fifteen separate booklets. Using the methods of EDA, it is designed to look at the role of statistics in the widest possible context in society.

Conventional statistics

There is no shortage of textbooks which deal with the conventional statistics that features so regularly in first year college courses. Here is just an indicative selection

Modern Business Statistics by Ronald L. Imam and W. J. Conover (John Wiley, New York, 1983).

My own favourite, it covers a few topics in EDA. It is well laced with real applications, and gives a very thorough, well presented description of statistics.

Statistics and Experimental Design in Engineering and the Physical Sciences (2 volumes) by N. L. Johnson and F. C. Leone (John Wiley, New York, 2nd ed. 1977).

For science or engineering applications, this is hard to beat, with clear explanation of the techniques.

Statistics for Social Scientists: An introductory text with computer applications by Louis Cohen and Michael Holliday (Harper & Row, New York, 1982).

Quantitative Methods for Business Students by Gordon Bancroft and George O'Sullivan London (McGraw-Hill, 1981).

Statistics Without Tears, A Primer for Non-mathematicians by Derek Rowantree (Penguin, Middlesex, England, 1981).

Theory and Problems of Statistics by Murray Spiegel, Schaum's Outline Series (McGraw-Hill, London, 1972).

A useful reference, which includes hundreds of worked examples. A good starting point if conventional techniques need to be applied or explained.

Discussion about statistics

There are a number of books, some quite readable, which deal with statistics in a general way. This could be on the role of statistics, or about some historical controversies. They expand the subject in an interesting way.

Ways and Means of Statistics by L. J. Tashman and K. R. Lamborn, (Harcourt Brace Jovanovich, New York, 1979).

Statistical Thinking. A Structural Approach by John L Phillips (W. H. Freeman, San Francisco, 2nd ed., 1982).

Statistics: a Guide to the Unknown by Judith Tanur (ed.) (Holden Day, San Francisco, 1978).

How to Lie with Statistics by Darrell Huff (Penguin, 1973)

17.3 STATISTICS ON THE COMPUTER

The programs you developed while completing the tasks in this book could be used for serious work. That was not the intention; they were learning tools, to enable you to try out ideas. If you are faced with a major statistical job, like analysing the results of a survey, you would want to make use of a standard commercially supplied program. So what is available?

(One problem with putting a list in a book is the speed with which it becomes out of date. Please check on the latest position before doing anything based on the information that follows.)

Micro or mainframe?

The advantages of the micro are well known to you as a micro user. It is handy, it is friendly, you control what goes on, it operates at your workplace. Since your main concern is the data, and what significance you can glean from the facts, the micro is ideal – you take it to the job, it sits quietly in the corner and is available when you need it.

The disadvantages of the micro for serious statistical work are mostly technical – unless you have disk drives, you are slowed down when loading up data or programs. Even with disks, you may have limitations on the amount of data you can handle. Perhaps most damaging of all is the fact that all the really good packages for statistical analysis have (to date) been developed for the mainframe.

The main advantage of the mainframe, in addition to being able to run all the best packages, is speed. The processor on the big computer will generally run many times faster, especially compared with an eight-bit micro. But this advantage is not such a big deal; unless you are operating on mammoth sets of data, the speed difference is only a matter of a few minutes. What may be more significant is the time you have to spend getting to a computer terminal plus the time you spend waiting for a terminal to become free.

Packages

SPSS (Statistical Package for the Social Siiences).

For many years this has been the most popular and most widely used package. It is backed by a US corporation, and is continuously developing, to adapt to changing customer requirements. The main criticisms of SPSS are: its clumsy command style (bearing traces of its Fortran punched card ancestry); and the fact that it was developed by sociologists and computer programmers and not statisticians. This package is widely available on mainframe computers in most universities, polytechnics and research establishments. It is well documented. Audio-visual learning aids are available as well. Up to now it has been available on mainframe computers only, but I am told (1985) that a version which will run on an IBM-PC will soon be issued.

Minitab

Minitab is rapidly gaining favour in educational establishments. In an exhaustive comparative survey by the Open University, it was selected as the 'best buy'. Unlike SPSS, Minitab was developed by statisticians for use by students learning statistics. It is very easy to use, with HELP available for all commands. It is interactive, with graphic limited only by the standard character set. As well as the usual statistical functions, Minitab includes time series analysis and exploratory data analysis. Minitab was designed as a mainframe package, but is also currently available in a CP/M microcomputer version. To run it, your micro has to be fairly powerful – twin double-sided disk drives, and 64K memory. But Minitab represents a real breakthrough – a proper statistics package, with full back-up on a microcomputer.

Other mainframe packages

SAS, BMMD, XDS3, GLIM, GENSTAT are the abbreviated names of some of the fifty or so packages which are available. GLIM and GENSTAT deserve special mention. They are both designed for more advanced statistical work. This means that their use is restricted to professionals, but they are very powerful packages.

Micro packages

There have been several piecemeal attempts to put together a number of statistical programs to create a 'package', but most were hampered in their development by the machine limitations.

One collection that I have found quite good is ASA – Advanced Statistical Analysis (Tandy), which is written in Microsoft BASIC, so should be available fairly widely.

A program which deserves special mention is EDA – (Apple), which makes imaginative use of the colour graphics, to produce box-plots, and stem and leaf plots.

Perhaps the most successful package ever developed for the micro-computer is Visicalc, which has spawned a whole clutch of 'visi-clones' – programs with similar spread sheet capabilities. You may not think of them as statistics packages, but they all feature excellent, friendly data input methods. This overcomes the poor input facilities of most existing micro and mainframe packages (Minitab has a rudimentary spread sheet approach). Within these spread-sheet packages it is possible to do quite a lot of simple statistical analysis. Latest developments give even more possibilities – we may be getting micro-statistics packages by this back-door method.

When using a statistics package you should avoid the temptation of using procedures which you do not fully understand. It is very easy to instruct the computer to perform elaborate calculations. The results may look impressive, but unless you know what they mean, the exercise is futile.

17.4 THE LONGER TERM

You may wish to deepen your knowledge by reading more about EDA or the other main area of statistics, confirmatory analysis. You may wish to find out more about using computer packages. All of this will help your understanding of what can be done, and what is meant when you read the results of other peoples' statistical work.

But statistics is not a spectator sport. By working through this book you have already gained both the skills and knowledge to do a great deal of statistical work; above all, I trust you have the *confidence* in your own ability to make use of what you have learned. Keep looking out for ways to use things like box-plots, or median polished tables. The more you use these techniques, the better you know them and the easier they become.

CHAPTER 18

WHAT IS DIFFERENT
ABOUT THIS BOOK?

This chapter is directed mainly at other teachers and lecturers, to explain some of the reasoning behind the approach to the subject taken in this book. It is a replacement for the conventional preface, which is, I believe, a distraction for the learner.

18.1 WHAT'S WRONG WITH CURRENT TEACHING OF STATISTICS?

For students who take statistics as a mainstream subject, there is not a lot wrong with most of the current syllabuses. Students on maths courses, students in sciences where statistics is a major subject, such as sociologists or educational psychologists use statistics a great deal, and appreciate the need to learn. Such students are a minority. For the great mass of students in business studies, economics, accounting, surveying, engineering, town planning, etc. the quick dose of statistics they receive, usually in their first year is not very popular or effective. It is seen as marginal or subsidiary to the main 'professional' subjects. Not surprisingly, statistics is neither understood, nor applied to anything like the extent it should be.

The causes for this sorry state are, I believe:

(1) Statistics as taught lacks relevance, or is given pseudo-relevance. Data which are handed out, sometimes dressed up with doubtful scenarios, are of little interest to the learner.
(2) Statistics is made difficult through needless mathematicising. Great effort is put into expounding formulas, and much work is directed to working through calculations.
(3) Too much stress is laid on statistics as scientific hypothesis testing. Most real statistical work is concerned with finding out; drawing inferences in a scientifically valid way will not come until later, if at all.
(4) Statistics syllabuses fail to take into account the opportunities offered by electronic calculators, let alone microcomputers.

(5) Teaching is generally at a single rate, with a linear model of the subject building up 'brick by brick'.

Although most students learn sufficient tricks to pass their examinations, the effect is shallow and shortlived. What is more persistent is the feeling of failure, the clear demonstration to the majority of their lack of ability. Above all, it is the damage done to the students' *self-confidence* in their own ability to apply statistics that I find most depressing.

But statistics is important! It should be learned, understood, used. It is the gateway to seeing things *as they are*. Many students appreciate this, especially later on, when they try their hand at a project.

18.2 WHAT SHOULD A COURSE IN STATISTICS CONTAIN?

For the majority of students who do not take statistics as a major subject, but need to be able to make use of some of the simpler techniques, a course which is more relevant, exciting, stimulating, enjoyable, mind-stretching is called for. In this book there are *five* main elements.

1. Exploratory data analysis (EDA)

EDA was developed by John Tukey, as a means of formalising what investigators had known for years was the first stage of a survey – finding out. 'It is important to understand what you *can do* before you learn to measure how well you seem to have done it', says Tukey. This exploratory approach is gaining acceptance as being much more relevant to the majority of students. Although hypothesis testing is the public face of statistics, 'proof' can only follow exploration. Topics such as index numbers or time series can be accepted for what they are – exploratory – and given their rightful place in the syllabus. After all, more professional statisticians deal with these exploratory topics than deal with the much vaunted, t, F and chi-squared tests.

2. Practical work of relevance

To fulfil the need to make this course both practical and relevant, learners collect their own data. Compare this with some courses which undertake so-called practical experiments of a purely classroom, contrived nature using artificial data. Others *give* the learner 'case-study' details. Neither can produce the same commitment the learner feels for his or her *own* data. It may be a little time consuming to collect real data, but that in itself is a learning feature. The benefits of real data of direct interest to the student cannot be underestimated.

The specific topic chosen for this book relates to housing, but lecturers and teachers are encouraged to substitute any alternative topic, so long as the data is interesting to the learner.

3. Computer enhanced learning

There are several fantasy views about the place of the computer in learning. The ultimate horror is CBT – Computer Based Teaching – which is a glorified form of the teaching machines of the 1960s. Such domineering devices may have a marginal role in teaching simple rule-following activity. CAL and CAI, computer-aided learning and instruction, allow the student more flexibility, but are limited in scope. They also suffer from poor graphics – output on a VDU is neither pleasant nor easy to read. By setting the parameters within which the student can operate, and by slavishly copying a linear model of learning (one thing after another) these machines could cripple the minds of a generation. Fortunately they are prone to break down, and may even be encouraged to do so by their victims.

Just because most of the efforts to use computers as an aid to instruction have, to date been unsatisfactory does not mean we should ignore them. The effect of computers is in some ways immediate whether the student has access or not. Much of the formula thrashing that passes for statistics teaching can and ought to be scrapped at once. But the computer opens up new opportunities to learn in different and exciting ways. If the learner could turn to a computer to try out a program however simple, how could that facility be used to advantage?

Formulas can be programmed of course; but the ease of repeated calculations means that experiments on data can be carried out in a way that was never before possible. In Task 9.3 for example, the student discovers that subtracting the mean from the sum of squares gives the smallest adjusted sum. Previously the teacher could either assert that it was true or could even prove it algebraically. Now we can let the learner *find out* with the aid of a microcomputer.

The home-based microcomputer offers the best possible environment for learning. The machine is small, informal, but above all under the control of the learner. I have successfully run this course using mainframe terminals and college micros, but personal micros would be better.

There are no complete programs in this book, only indications of how the programs should be written. This has one practical advantage – the book is not machine dependent. It also creates an intellectual challenge in program writing. My experience has been that this is often the first time that learners get to grips with BASIC.

4. Self-paced independent learning

The course is designed to be used entirely independently of either classroom or teacher. Control of pace of learning, where the learning takes place, and at what times can be a matter for the learner to decide. Learning in a class can have advantages, not least the opportunity to discuss difficulties, and compare conclusions. It is still helpful if the learner can get help from an instructor in a classroom situation. The rigorous self-discipline of learning totally alone is not a gift that everyone possesses. An encouraging word, a timely reminder, a second opinion can get the learner past a stumbling block.

If an assessment of the student's performance is required, may I suggest that a completed notebook, detailing the student's success at finishing the tasks could be used for 'scoring'. Marks could be awarded for completeness – having done all the tasks, plus marks for layout, intelligent comment on the results of the tasks, effectiveness of the programs.

5. Activity and experimentatation

On this course the learner is not a passive observer, merely reading the text. The physical activity of collecting the data is the basis for further analysis. Interacting with the computer also involves activity. But it is much more than that: it is a form of manual labour. This is beneficial in engaging the whole brain in the learning process. The left-brain logic aspects of algebra are dealt with in conventional book-based courses. In this book, through activity, the right-brain aspects of imagination, intuition and feeling are brought into play as well.

18.3 SELF-CONFIDENCE

Completing the Tasks in this book will certainly give the student the skill to use statistics and to write computer programs. From practical experience with real data, students will have learned how to turn raw data into information and understanding. But above all else, by completing step-by-step achievable Tasks, learners develop *confidence* – the self-confidence that they have Mastered Statistics and *can* explore their own data for their own purposes.

INDEX